Water Resources Development and Management

Indexed by Scopus

Each book of this multidisciplinary series covers a critical or emerging water issue. Authors and contributors are leading experts of international repute. The readers of the series will be professionals from different disciplines and development sectors from different parts of the world. They will include civil engineers, economists, geographers, geoscientists, sociologists, lawyers, environmental scientists and biologists. The books will be of direct interest to universities, research institutions, private and public sector institutions, international organisations and NGOs. In addition, all the books will be standard reference books for the water and the associated resource sectors.

More information about this series at https://link.springer.com/bookseries/7009

M. Dinesh Kumar · Nitin Bassi · Saurabh Kumar

Drinking Water Security in Rural India

Dynamics, Influencing Factors, and Improvement Strategy

M. Dinesh Kumar
Institute for Resource Analysis and Policy
Hyderabad, Telangana, India

Nitin Bassi
Institute for Resource Analysis and Policy
Rohini, Delhi, India

Saurabh Kumar
Institute for Resource Analysis and Policy
Hyderabad, Telangana, India

ISSN 1614-810X ISSN 2198-316X (electronic)
Water Resources Development and Management
ISBN 978-981-16-9200-0 ISBN 978-981-16-9198-0 (eBook)
https://doi.org/10.1007/978-981-16-9198-0

This Springer imprint is published by the registered company Springer Nature Singapore Pte Ltd.
The registered company address is: 152 Beach Road, #21-01/04 Gateway East, Singapore 189721, Singapore

Preface and Acknowledgements

More than seven decades have passed since India got its independence. Yet, the country is struggling to make sure that every household has secured water supply for human uses in terms of quantity, quality, reliability that it can access with ease. Though the country has made tremendous strides in providing safe drinking water to rural areas through formal sources, a small fraction (18%) of the rural population has access to treated water for domestic uses within the dwelling. The rest has to access water from the public drinking water source located near the dwelling or away from the premise. A small but considerable proportion of the people in rural areas still depend on water from unprotected public water supply sources (like open wells), and a smaller fraction depend on water from unsafe sources like ponds and tanks.

Rural drinking water security means that the rural households should have year-round access to treated water within the dwelling premise in adequate quantities. A major problem hindering the process of achieving a hundred percent coverage of formal public supply of treated water to individual households is the dependability of the source. Over the past two to three decades, India had invested heavily in building village and hamlet-based mini water supply schemes based on groundwater, without looking at the probability of obtaining good-quality water throughout the year. A small fraction of the rural households gets water from schemes that are based on surface reservoirs even in areas where surface water is available in plenty. Most importantly, adequate attention has not been paid to the dynamics of change in groundwater availability across years and seasons and to the issue of competition that rural water supply schemes face from the irrigation sector. The incidence of failure of such schemes is very high, especially during the peak summer season.

The Comptroller and Auditor General (CAG) of India's performance audit report (2018) of the National Rural Drinking Water Project noted: 'coverage of rural habitations increased by only 8 per cent at 40 lpcd (litres per capita per day or water available per person) and 5.5 per cent on the basis of 55 lpcd during 2012–17 despite the expenditure of Rs. 81,168 crore'. Billions of dollars are spent every year by the state governments in India to provide tanker water to villages during peak summer months. According to the CAG report, as of December 2017, only 44 per cent of rural habitations and 85 per cent of government schools and anganwadis could be provided access to safe drinking water, only 18 per cent of the rural population could be provided with potable water through piped water supply (at 55 lpcd) and 17 per cent of the rural households with household connections.

Yet, there has been no systematic attempts to identify the factors responsible for the failure of groundwater-based schemes (wells, bore wells and handpumps) and identify the parameters that influence the performance of rural water supply schemes positively and negatively, to find out what types of schemes work under what conditions. Even after the launching of the ambitious programme of Jal Jeevan Mission, which envisioned piped water supply to every Indian household, by the Prime Minister of India in June 2019, the state governments continue to invest money on wells and bore wells and look at artificial recharge schemes as a way to augment them when they fail. Despite the poor effectiveness of artificial recharge schemes in most areas where wells fail, there is a clamour for them from most states. A plethora of fashionable concepts was introduced by the Department of Drinking Water and Sanitation of the government of India as the key to address the issue of sustainability. Some of them are: 'water budgeting', 'village-level water safety and security planning', 'source strengthening', 'creating awareness of water conservation', 'capacity building of community organizations', etc. However, they practically meant nothing in terms of changes on the ground.

On the other hand, the increasing evidence of greater sustainability of water supply schemes that are based on surface reservoirs, and better adoption of household tap connections and improved toilets by households that are covered by such schemes are largely ignored.

In this book, we synthesize the results and findings of studies carried out over the past 3–4 years by the Institute for Resource Analysis and Policy (IRAP) on rural water supply, and rural water safety and security in India. The studies investigated the determinants of performance of rural water supply schemes, identified the conditions under which groundwater-based schemes are unlikely to perform well, and evolved long-term strategies for improving the sustainability of rural water supply schemes in different regions of India where groundwater-based schemes show high incidences of failure either due to availability related issues or water quality issues.

We have used several quantitative and modelling techniques to explain: differential groundwater behaviour during monsoon across physical settings; and differential performance of water supply schemes across physical and socioeconomic settings. We also employed innovative methods to quantify un-committed surface water resources in river basins, crucial for ascertaining the viability of an alternative strategy for rural drinking water security, by using Maharashtra River Basins as a case. We also show how the available reservoir storage or un-utilized water resources in different river basins and regions of India can be used for augmenting rural water supply within the region or outside where sustainability is still an issue.

Moving from water quantity management to water quality management, in this book, we have also discussed the concept of drinking water quality surveillance and presented the development of a composite index that helps identify the regions or pockets that are susceptible to water pollution and require intensive monitoring of source water to protect drinking water sources. It also presents the computed values of the index for each block of Maharashtra to identify the areas that require surveillance and monitoring to ensure safety of drinking water sources.

We have also discussed the institutional and policy reforms that would be required to bring about changes in the current government outlook on rural drinking water security, which is heavily driven by considerations of low capital investment for the infrastructure and decentralized management. We have also done a rapid quantification of the economic benefits that can be accrued from infrastructural investments for improved rural water supply against the costs that are likely to be involved. We sincerely hope the analyses and arguments presented in this book will be of value to water supply professionals, policy makers in the rural water supply sector, and academics who are working on groundwater dynamics and rural drinking water security.

For developing this book, we have extensively used the research materials and outputs developed by IRAP at various points of time with the support of UNICEF offices in India, apart from the work done by IRAP researchers in the past. While we have acknowledged them properly at appropriate places in the book, we would also like to express our gratitude to UNICEF for their continued support. The authors are extremely grateful to Ms. Vaishnavi Potharaju, Research Officer-IRAP for her support in extensively reviewing published literature and preparing notes that were immensely helpful in drafting certain chapters of the manuscript. We are thankful to Ms. Archana Dineshkumar Manhachery, Consultant-Science Editing, for meticulously editing the manuscript for scientific accuracy, consistency, language, and readability. We are also highly thankful to Mr. Ajath Sanjeev, Executive Assistant to Director-IRAP, for rigorously carrying out the reference check and preparing the 'Table of Content' and 'Abbreviations' for the manuscript.

Hyderabad, India M. Dinesh Kumar
Rohini, India Nitin Bassi
Hyderabad, India Saurabh Kumar

Contents

Abbreviations

AP	Andhra Pradesh
ARWSP	Accelerated Rural Water Supply Programme
ATSDR	Agency for Toxic Substances and Disease Registry
BCM	Billion cubic metres
BHC	Benzene hexachloride
BIS	Bureau of Indian Standards
BOD	Biochemical oxygen demand
CCA	Cultivable command area
CGWB	Central Ground Water Board
COD	Chemical oxygen demand
CPCB	Central Pollution Control Board
CR	Commissioned Report
CRI	Compliance Risk Index
CTARA	Centre for Technology Alternatives for Rural Areas
CV	Coefficient of variation
DDT	Dichloro-diphenyl-trichloroethane
DNA	Deoxyribonucleic acid
DO	Dissolved oxygen
DW	Drinking water
DWI	Drinking Water Index
DWQSI	Drinking Water Quality Surveillance Index
EFR	East flowing rivers
FAO	Food and Agriculture Organization
FHTC	Functional household tap connection
FRL	Full reservoir levels
GIS	Geographic Information System
GOI	Government of India
GP	Gram Panchayat
GPCB	Gujarat State Pollution Control Board
GPS	Global Positioning System
GSDA	Groundwater Surveys and Development Agency

GW	Ground water
GWSSB	Gujarat Water Supply and Sewerage Board
HH	Households
HP	Hand pumps
IGNP	Indira Gandhi Nahar Project
IMD	India Meteorological Department
INR	Indian rupee
IRAP	Institute for Resource Analysis and Policy
IWMI	International Water Management Institute
JJM	Jal Jeevan Mission
LCCA	Life-cycle cost approach
MCM	Million cubic metre
MD	Mekong Delta
MGNREGS	Mahatma Gandhi National Rural Employment Guarantee Scheme
MJP	Maharashtra Jeevan Pradhikaran
MNP	Minimum needs programme
MODWS	Ministry of Drinking Water and Sanitation
MPCB	Maharashtra Pollution Control Board
MWRRA	Maharashtra Water Resources Regulatory Authority
NCT	National Capital Territory
NFHS	National Family Health Survey
NIDM	National Institute for Disaster Management
NIH	National Institute for Hydrology
NRDWP	National Rural Drinking Water Programme
NREGS	National Rural Employment Guarantee Scheme
NWDA	National Water Development Agency
NWDT	Narmada Water Dispute Tribunal
ODF	Open defecation free
PHED	Public Health Engineering Department
PM	Prime Minister
PWS	Piped water supply
RL	River lift
RNA	Ribonucleic acid
RO	Reverse osmosis
RWS	Rural water supply
RWSS	Rural water supply services
SBM	Swachh Bharat Mission
SCADA	Supervisory Control and Data Acquisition
SDP	State domestic product
SRP	Sector Reforms Project
SSNNL	Sardar Sarovar Narmada Nigam Limited
SSP	Sardar Sarovar Project
SW	Source water
TDS	Total dissolved solids
TMC	Thousand million cubic

UK	United Kingdom
UMA	Unnat Maharashtra Abhiyan
UNICEF	United Nations International Children's Emergency Fund
UP	Uttar Pradesh
USA	United States of America
USEPA	United States Environmental Protection Agency
UT	Union territories
VWSC	Village Water and Sanitation Committee
WASH	Water, sanitation and hygiene
WATSAN	Water supply and sanitation
WB	West Bengal
WELL	Water and Environmental Health at London and Loughborough
WFR	West flowing rivers
WHO	World Health Organization
WLF	Water level fluctuations
WQ	Water quality
WQM	Water quality monitoring
WQSI	Water Quality Surveillance Index
WRD	Water Resources Department
WSP	Water and Sanitation Programme
WSSO	Water and Sanitation Support Organization
WTF	Water-table fluctuation

List of Figures

List of Tables

Chapter 1
Rural Domestic Water Supply in India: Progress and Issues

1.1 Introduction

At the current population size, India needs to supply nearly 22 billion m^3 of water annually to rural areas for the rural population to satisfy their basic water needs. While the country has the second largest irrigated area in the world, with an annual total water withdrawal exceeding 600 BCM, a significant proportion of India's rural population experiences stress in managing sufficient water of adequate quality to meet their basic needs of drinking, cooking, and personal hygiene during summer months. Ever since independence, India had made a whopping investment for improving water supplies in rural areas through various schemes and programmes implemented during the various five-year plans up to 2017 and thereafter through annual budgetary allocations. During the last decade (2010–2020) alone, INR 1373 billion was spent (Source: Department of Drinking Water and Sanitation, Ministry of Jal Shakti, Government of India). The aim underlying these efforts was to improve the coverage of the schemes in terms of the number of villages and habitations and supply potable water to rural areas. Starting in , the focus shifted to providing individual household connections.

Since the 1950s, the rural domestic water supply sector has witnessed a series of policy and technological and institutional reforms. As part of the First Five-Year Plan (1951), the national water supply and sanitation programme was started. During 1956–72, the Government of India (GoI) allotted resources to state governments for strengthening and equipping the Public Health Engineering Department (PHED) to undertake rural water supply schemes which were extended to small urban towns and villages. During this period, GoI introduced Accelerated Rural Water Supply Programme (ARWSP) which was changed and incorporated into the minimum needs programme (MNP) in 1974. The latter was soon abandoned, and ARWSP was revived in 1977. Till the mid of Ninth Five-Year Plan (1999), approach to the provision of rural water supply in India was supply-driven with an emphasis on norms and targets and on the construction and creation of assets, with relatively little concern for suitable arrangements for the less attractive but critical issue of better management and

M. Dinesh Kumar et al., *Drinking Water Security in Rural India*, Water Resources Development and Management, https://doi.org/10.1007/978-981-16-9198-0_1

maintenance of the facilities already built (WSP 2004). The water supply requirements of the scheme were decided based on certain simplistic norms on the per capita water requirements of the rural households and the population to be covered. Consequentially, no attention was paid to the socio-economic characteristics of the population to be served, the climatic conditions, and cultural aspects which would have a major bearing on the actual water needs in rural areas for domestic purposes (Bassi et al. 2021). Further, the productive water needs were completely ignored (Kumar et al. 2016).

Moreover, the efforts were only on increasing coverage with little concern for the sustainability of the technical system created and the resource base which these systems tapped (Reddy et al. 2010). Overdependence on groundwater-based schemes resulted in drying up of the sources such as open wells, tube wells, and handpumps as a result of over-exploitation of groundwater and drop in water levels. Because of this poor vision on water resources management and operation and maintenance (O&M) of the systems, at several places, rural water supply infrastructure rarely lasted its design life and its service quality was far below the expectation. Inefficient service delivery, poor quality of infrastructure, and poor maintenance of the system under the supply-driven approach led to a fundamental policy shift in the sector by the end of the 1990s, towards an approach, which was widely acclaimed as 'demand-driven approach' to rural water supply. Under the new approach, community participation and decentralization of powers for implementing and operating drinking water supply schemes was made a prerequisite. It was envisaged that the role of the government would be that of a facilitator. This was to be achieved through the Sector Reforms Project (SRP) which was launched in the year 1999 on a pilot basis in 67 districts across the country.

By 2003, the SRP was scaled up to cover the entire country under the *Swajaldhara* Programme. *Swajaldhara* guidelines envisaged 10% of the capital cost and 100% of the O&M cost to be borne by beneficiaries, with the central government providing the remaining. The focus continued on the development and promotion of groundwater-based schemes. However, the limited external assessments concerning the implementation of the Swajaldhara scheme showed that there were serious pitfalls in the proposed institutional reform principles (Sampat 2007). The government agencies responsible for the implementation of the drinking water supply policy were not keen on bringing about the changes that are required to hand over some of the powers, like planning and operation of drinking water schemes, to local communities. Further, the principle of seeking community contribution towards capital cost was politically unappealing for the state governments (Cullet 2011). As a result, in many villages, Swajaldhara projects did not follow the guidelines.

During the 11th Five-Year Plan (2007–2012), a new policy instrument, i.e. National Rural Drinking Water Programme (NRDWP), was formulated. The objective of NRDWP was to provide safe and adequate water for drinking, cooking, and other domestic needs to every rural person on a sustainable basis. Further, NRDWP emphasized the need to conceive drinking water supply in a wider context of public health and proposed its linkage with the Mahatma Gandhi National Rural Employment Guarantee Scheme (MGNREGS) (Cullet 2011). However, attempts to link with other programmes such as MGNREGS had undesirable consequences. MGNREGS is strictly an employment guarantee scheme that aims at providing 100 days of unskilled manual work to every rural household. Its role in building water supply schemes, which require both skilled labour and engineering infrastructure, is too meagre (Bassi et al. 2014a). In many instances, this led to local government agencies undertaking de-silting of ponds, tanks, and check dams in the name of 'water supply improvement or augmentation', to tap MGNREGS funds for the same (Bassi et al. 2014a). However, it did little to improve the sustainability of the water supply scheme or the water source. NRDWP continued till , and even after a decade, it failed to provide access to safe drinking water to all the habitations, government schools, and day care centres in the rural areas. Further, many new water supply schemes became non-functional as the focus was on groundwater-based schemes which could not provide year-round water (GoI 2018).

In , the Government of India launched Jal Jeevan Mission (JJM) intending to provide safe and adequate drinking water through functional household tap connection (FHTC) to every rural household by 2024. By the end of July 2021, 78 million rural households, which are about 41% of the total rural households, were provided with FHTC. Though, since the start of the JJM, the proportion of households with FHTC have increased by 2.5 times, there are still about 113 million households without FHTC. Providing a functional household tap connection has two aspects: (1) providing the pipeline and taps within the dwelling premise by extending the existing water distribution system in the villages and (2) providing water supply round the year of sufficient quantity and quality, from a dependable source. It is an established fact that unless the households are convinced of getting adequate quantities of water throughout the year for meeting their domestic needs from their domestic tap connection, they would neither invest in building the additional physical infrastructure nor pay for the water supply services. The strategy of providing a functional household tap connection should therefore be built on a sound water resources management strategy.

Given the continued emphasis on the failed model of building 'single-village' schemes based on groundwater and promoting decentralized management of the schemes by the Gram Panchayats, it is important to know whether the technical strategy adopted for achieving this ambitious target would work in the current scenario or not—especially in the light of the growing competition for groundwater from the irrigation sector and the lack of any effective regulation to check groundwater abstraction. An objective assessment of the situation on the ground vis-à-vis the sustainability of different types of water supply schemes shows that the single-village schemes based on groundwater are not sustainable. The government programmes to strengthen such schemes through recharge structures have been totally ineffective. Overall, the inputs for planning rural drinking water supply through a 'decentralized model' have been devoid of any scientific assessment of groundwater resource dynamics in response to recharge, abstraction, and natural processes and the factors influencing the performance of drinking water wells.

1.2 Progress in Rural Water Supply Made by Indian States and Union Territories

A remarkable achievement has been made in improving the coverage of water supply in terms of the number of habitations in the rural areas. By the end of April 2021, 79% of all the rural habitations were covered by schemes that were designed to supply 40 lpcd (litres per capita per day) of water. However, 22 out of the total 33 states and union territories (UT) have water supply coverage below the national average. In only four states, i.e. Goa, Gujarat Madhya Pradesh, and Telangana, all the rural habitations have an average of 40 lpcd of water supply. The state of Meghalaya, which receives the highest quantum of rainfall annually, had the lowest proportion of rural habitations (26%) covered by formal water supply from public water supply schemes. Overall, the coverage was lowest in the north-eastern Indian states and was highest in the central Indian states (Table 1.1). A large portion of the north-east has mountainous terrain. Improving formal water supply coverage in such areas is prohibitively expensive and challenging as gravity-based schemes do not work in most cases, and substantial costs will have to be incurred for lifting water from streams to areas at high altitudes.

In terms of the scheme type, groundwater-based schemes are predominant. Almost 90% of the total rural water supply schemes are based on groundwater. In the north-eastern and north-western regions, the proportion of schemes that are groundwater based is less than the national average. This is because of the abundance of water in streams and springs on the one hand and the extreme difficulty in obtaining water

Table 1.1 Proportion of rural habitations covered with schemes designed to supply 40 lpcd of water

Indian regions	States/UTs	% rural habitations covered
North-east	Arunachal Pradesh, Assam, Manipur, Meghalaya, Mizoram, Nagaland, Sikkim, Tripura	46
West	Daman and Diu, Dadra and Nagar Haveli, Goa, Gujarat, Maharashtra, Rajasthan	65
South	Andaman and Nicobar, Andhra Pradesh, Karnataka, Kerala, Puducherry, Tamil Nadu, Telangana	66
North-west	Haryana, Punjab	84
East	Bihar, Jharkhand, Odisha, West Bengal	85
North	Himachal Pradesh, Jammu and Kashmir, Ladakh, Uttarakhand, Uttar Pradesh	89
Central	Chhattisgarh, Madhya Pradesh	99
Overall		79

Source Authors' analysis using data from the Department from Drinking Water and Sanitation, Government of India from wells in the hilly tracts on the other hand. While in the case of the north east, communities depend on local water sources such as ponds and springs, in the case of the north-west, they take advantage of the extensive network of canals for meeting domestic water demands. In the remaining five regions, more than 90% of the schemes are based on groundwater. The remaining five regions that are heavily dependent on groundwater-based schemes are largely underlain by consolidated rocks (refer to Table 1.2). These formations have poor storage potential due to low specific yield. Since irrigation water demand is excessively high in these regions, especially after the monsoon owing to relatively low rainfall, high aridity and the presence of large amount of arable land, due to high dependence on groundwater for irrigation, wells go dry in such areas even before the onset of summer months leading to the failure of wells. In dry years, the situation becomes even worse when there is an acute water shortage and the government resorts to supplying drinking water from distant sources through tankers.

Table 1.2 Proportion of groundwater-based schemes and rock formations in India

Indian regions	States/UTs	% GW-based schemes	Major rock formations
North	Himachal Pradesh, Jammu and Kashmir, Ladakh, Uttarakhand, Uttar Pradesh	99	Consolidated to unconsolidated and hilly areas
North-east	Arunachal Pradesh, Assam, Manipur, Meghalaya, Mizoram, Nagaland, Sikkim, Tripura	84	Unconsolidated and crystalline formations
West	Daman and Diu, Dadra and Nagar Haveli, Goa, Gujarat, Maharashtra, Rajasthan	91	Consolidated, semi-consolidated, and unconsolidated
North-west	Haryana, Punjab	75	Unconsolidated
Central	Chhattisgarh, Madhya Pradesh	100	Mostly consolidated
East	Bihar, Jharkhand, Odisha, West Bengal	100	Consolidated (Jharkhand and Odisha) to unconsolidated (Bihar, West Bengal)
South	Andaman and Nicobar, Andhra Pradesh, Karnataka, Kerala, Puducherry, Tamil Nadu, Telangana	96	Consolidated, with patches of unconsolidated formations along the coast
Overall		97	

Source Authors' analysis using data from the Department of Drinking Water and Sanitation, and Central Ground Water Board, Government of India

Most groundwater-based schemes are designed for 40 lpcd and surface water-based schemes for 55 lpcd. The current water supply norms miss out on two important factors: (1) local climate and (2) the socio-economic (for domestic, livestock, kitchen garden, etc., uses) water demand of the households (Bassi et al. 2021). However, there is now a growing recognition that in the hot and arid regions the per capita water requirements for domestic uses like animal watering and bathing are even higher than that in cold and humid regions, and therefore, more water needs to be supplied for domestic uses in such areas (IRAP et al. 2018). On the socio-economic front, the water demand for domestic purposes has increased remarkably owing to rising income levels, improvement in lifestyles, and changing occupational patterns (Bassi et al. 2021). Thus, it can be concluded that the current water supply norms are inadequate to meet the water demand of rural households.

Further, as of April 2021, in about three per cent of rural habitations (48,000 in total), the quality of source water (mostly groundwater) is poor and unfit for drinking (Source: Department of Drinking Water and Sanitation, Ministry of Jal Shakti, Government of India). The Assam, Bihar, Odisha, Punjab, Rajasthan, Tripura, and West Bengal states are the worst affected. Assam and Rajasthan together constitute about 70% of water quality-affected habitations. While the former experiences frequent floods resulting in high bacteriological contamination in groundwater, the latter is among the most arid regions of the world and has high salinity and fluoride concentration in groundwater.

1.3 Threats to Sustainability of Rural Domestic Water Supply

The overall achievements with regard to improving domestic water supply do not commensurate with the scale of investments that have gone into the sector. Though coverage has improved, progress in terms of functional piped water supply to rural households has been slow. Moreover, the water supplied as per the current norms is inadequate to meet the household water demand, and in several cases, the water source is unfit for drinking. Further, an increasing number of habitations are slipping back to the 'no-source category'. This is because of the several threats rural water supply sector is facing, which include poor sustainability of water supplies owing to unsustainable resource base (groundwater); soaring operation and maintenance costs due to the absence of regular upkeep, and poor cost recovery, both resulting in the poor financial working of the system; and inequity in access to water across different segments owing to lack of adequate institutional capacities built at the local level for water distribution (Kumar et al. 2016). These threats are discussed in detail below.

1.3.1 Technical Sustainability of Groundwater-Based Schemes

A large part of India, about 2/3 of the landmass, is covered by unconsolidated or hard rock formations. In a large proportion of these regions (barring the hills of the north-east and Western ghats), rural water supply schemes tap groundwater, while the schemes in the unconsolidated formations (especially in the Indo-Gangetic Plains) are more heavily dependent on groundwater. Groundwater behaviour in hard rock formations is complex due to a number of factors influencing the water balance. They include natural discharge into streams, rejected recharge due to limited storage space, etc. Hence, it is difficult to derive any meaningful inference related to the availability of water in such aquifers purely based on analysis of the groundwater-level trends at the aggregate level. A more nuanced understanding of the aquifer characteristics and its response to rainfall and pumping is required. During the monsoon months, the water level rises continuously in such aquifers due to very low specific yield. The small quantity of water pumped for domestic uses does not cause any decline in the water table during the season, as it is insignificant in comparison with the recharge effect of rainfall.

It is common to find overflowing dug wells in the hilly and undulating regions during monsoon, as the cumulative infiltration from rainfall exceeds the storage potential of the aquifer (space in weathered formation and the fissures) (Kumar et al. 2006). However, once the monsoon ends, there is a sudden fall in groundwater level, usually starting in November, as pumping for irrigation of winter crops starts and there is a very limited amount of recharge during non-monsoon months, unlike in canal command areas where return flows augment groundwater recharge during the lean season. The water abstraction for irrigation peaks during December and January, when the soil moisture gets exhausted. Though the irrigation demand comes down during the summer months, the aquifer no longer has any water left, thus leading to seasonal water scarcity. In addition to the abstraction, there is also the natural discharge of water from the formations, mainly influenced by the geomorphology. While in normal rainfall years, there would be hardly any water to meet the irrigation demand of summer crops grown in tiny plots, during dry years, there is a shortage of water to meet the domestic needs during the season. Thus, in such formations, groundwater is not a reliable source for domestic water supply.

1.3.2 Financial Sustainability of the Schemes

There are serious challenges for the effective O&M of the existing water supply schemes. Most of the groundwater-based schemes and single-village surface water-based schemes are handed over to the local community for upkeep. A village water and sanitation committee (VWSC), a subcommittee of the village *panchayat*, is entrusted with the responsibility of O&M of the scheme. However, as a study in

Maharashtra has shown, most VWSCs lack technical, managerial, and financial capabilities for proper O&M of the system (Bassi and Kabir 2016). As a result, in most of the states, O&M plans of the water supply schemes are either not prepared or have deficiencies to effectively deal in case of the scheme breakdown (GoI 2018). Further, in many states, water for domestic use is supplied with heavy subsidies. Often, funds are not available with VWSC for proper upkeep of the system as the recovery of even subsidized water charges, especially for the groundwater-based scheme, is low (Bassi and Kabir 2016). This results in poor performance of such schemes, more so during summer months when they are unable to provide the required quantity of water. Thus, a vicious cycle is formed comprising subsidized water supply, low recovery of water charges, poor O&M of system, frequent breakdown, and low willingness of the community to pay for water, leading to poor financial sustainability and physical performance of the scheme.

1.3.3 Inequity in Access to Water

Domestic water requirement varies as per the socio-economic condition of the households. Water requirement for rural households with high per capita income would be high as their living standards improve and they adopt improved sanitation technologies and acquire assets such as washing machines, water filters, heating rods, etc., providing them with more convenience. Also, with the increasing number of households in water-scarce regions opting for intensive dairy farming, the demand for water for watering animals has increased (FAO 2006). Fetching water for livestock is as big a priority for dairy farmers as finding water for meeting other domestic purposes. Further, rural households may have other productive water needs including for kitchen gardens, homesteads, livestock keeping, and running small-scale industries (Kumar et al. 2016).

Studies have shown that with the rise in income, the proportion of households having access to piped water connections within the dwelling premise goes up (Jalan and Ravallion 2003), leading to an increase in per capita domestic water consumption considerably (WELL 1998; Howard and Bartram 2003). Further, such households invest in developing their private water sources such as dug wells and handpumps, thus reducing dependence on public water supply which is irregular and provides them with an option of accessing multiple water sources (Bassi et al. 2021). However, households with low incomes mostly depend on distant public water supply sources or purchased water for meeting their water demand. Thus, well-off households have better access to water than others.

Nevertheless, studies have found that surface water-based water supply schemes fare better than the groundwater-based schemes vis-a-vis equity in the distribution of water (Bassi et al. 2014b; Bassi and Kabir 2016). They tend to perform better in terms of coverage, daily hours of water supply, and access to water within the premise. This is irrespective of the socio-economic conditions of the households.

1.4 Rationale for the Book

It is clear from the discussions in the previous sections that the groundwater-based schemes that dominate the rural water supply sector are underperforming. With a major portion of such schemes being located in the hard rock areas where aquifers do not yield sufficient water to provide year-round supply, such schemes will continue to fail. Yet, there is no visible shift in policy from investing in single-village schemes based on bore wells to reservoir-based schemes that are easy to manage with regard to controlling the use of water from the source and physical allocation of water. The operational guidelines for JJM, which aims to provide all rural households with functional tap connection, focus mainly on strengthening the existing groundwater-based village schemes, using rainwater harvesting, watershed development, groundwater recharge, and de-silting and storage capacity enhancement of water bodies (GOI). They are based on the wisdom on what has supposedly worked in the past in tackling drinking water scarcity in the country and/or the information available in the public domain on what can work in the future, combined with the assumption that farmers will prioritize drinking water over irrigation in their allocation decisions. However, this is far from the truth.

In reality, many semiarid and arid regions with low to medium rainfall and underlying hard rock formations face seasonal scarcity of groundwater for irrigation and drinking water. In such regions, there is large unmet demand for groundwater in irrigation given the vast areas of arable land, and limited access to surface irrigation. Further, groundwater recharge is also poor in such regions. Though the official estimates of groundwater development for such regions show high annual utilizable recharge, this is an anomaly due to the inaccurate methodology followed for assessing the status of groundwater development (Kumar and Singh 2008). They tend to over-estimate the utilizable groundwater recharge and probably under-estimate abstraction. Further, the demand for groundwater in irrigation in these regions is mainly during winter and to some extent during monsoon months (depending on rainfall conditions). As a result, even in good rainfall years, the water available in the wells is pumped out for irrigation during winter itself, leaving no water for summer, and as a result, drinking water schemes fail. Large-scale investments are made year after year for drilling new wells and handpumps, with a high percentage of 'slippage', as these sources go dry during summer months and completely become dysfunctional during droughts.

Clearly, the assumptions on which the options have been identified for improving the sustainability of existing schemes need a relook. Overall, it is evident that the JJM guidelines have inherent limitations in capturing the vast heterogeneity in the physical environment (hydrology, geohydrology, topography, climate, etc.), socio-economic dynamic (water demands from other sectors, especially agriculture in rural areas, demand pattern, and economic and social conditions of the rural people) and institutional capacity available at the local level for managing water that governs the availability of and access to water for drinking purpose in rural areas. Water, being a state subject in India, the responsibility to contextualize it rests with the state government and the implementing departments. Even if the GPs or the VWSCs are to do the operation and maintenance of the village-level schemes, their technical capacities need to be enhanced.

There has been published scientific literature on the outcomes of the policy reforms in the rural water supply sector in India. The major focus was on the techno-institutional models adopted for implementing the reforms (Sangameswaran 2010; Bassi et al. 2014b) and evaluation of the demand-led, participatory, and democratic decision-making process in the implementation of the water supply project (Prokopy 2005; Brunner et al. 2010). Some other studies have explored the various domestic and productive water demands of the rural community in India and pitched for designing rural water supply systems catering to multiple water uses (Smits et al. 2010; Kumar et al. 2016). Also, climate-induced risks affecting the proper functioning of the rural water supply schemes have been assessed (Howard and Bartram 2010; Kumar et al. 2021) and strategies to make them resilient have been proposed (Kumar et al. 2021). While performance assessment of different techno-institutional models of rural water supply had shown the better performance of reservoir-based schemes when compared to schemes based on wells in hard rock areas (see Bassi and Kabir 2016), there is no analysis of the critical factors influencing the performance of groundwater-based schemes.

While the national and state governments continue to invest in building water supply schemes based on groundwater sources and improving their sustainability, the inputs for planning large-scale investments for such schemes are devoid of any scientific understanding of the groundwater resource condition and behaviour and the factors influencing the performance of drinking water wells in typical rural settings. It is largely driven by the ideology that rural drinking water schemes should be managed by local village communities for them to be effective, and it is much easier for them to manage small, decentralized schemes. In fact, over the past three and a half decades, community management has become the accepted management model for rural water supply (RWS) in low- and middle-income countries (Schouten and Moriarty 2003; Harvey and Reed 2007; Lockwood and Smits 2011).

While the world has made great strides expanding access to improved water supply, with a further 25% of the global population gaining access between 1995 and 2010 (UNICEF and WHO 2012), evidence from both India (Reddy et al. 2010) and sub-Saharan Africa (Baumann 2006) shows that around a third of RWS systems are non-functional, raising serious questions about the effectiveness and sustainability of the community management model. Moriarty et al. (2013, p. 329) argue we are now at the 'beginning of the end' for community management, 'not principally because community management has failed, but because it is reaching the limits of what can be realistically achieved in an approach based on informality and voluntarism'. From a policy perspective, the approach falls short in two main areas: lack of long-term sustainability and lack of scalability across large projects (Bolt et al. 2001).

However, the focus of the research on the effectiveness of community-managed schemes has been on the institutional effectiveness (see Hutchings et al. 2015), and not on the influence of technology choices on the scheme's effectiveness and sustainability. But the wrong choice of the source water and technology for water extraction, distribution, and delivery can influence the effectiveness and sustainability of the scheme. It is important to understand the dynamics of the resource that the schemes tap and the factors that influence the performance of the schemes, particularly under what conditions they succeed and when they fail, other than merely looking at the institutional effectiveness.

1.5 Objectives and Scope of the Book

The main objectives of the book were to: deepen the understanding of groundwater behaviour in response to natural processes in different geological settings; analyse the factors influencing the performance of rural water supply schemes; identify the conditions under which groundwater-based drinking water sources become sustainable and suggest measures for improving the sustainability of drinking water wells in hard rock regions that constitute nearly two-third of India's geographical area, and finally suggest physical strategies and policy measures for improving the long-term sustainability of rural drinking supply schemes in India.

For this, first, statistical models were developed and validated using time series data of groundwater levels (pre-monsoon depth to water levels, water level fluctuations during monsoon), rainfall, the recharge potential of the surface formations and aquifer storage space to explain groundwater behaviour under different geological settings from the state of Maharashtra, as the state represents a microcosm of the various hydrological, topographical, and geohydrological conditions that are encountered in the country. Statistical models were also developed to explain the varying degree of success and failure of rural water supply schemes and validated using district-level data on all the above-mentioned parameters, plus data on geomorphology, extent of surface irrigation, extent of dependence on groundwater-based schemes for water supply, irrigation demand per unit sown area under different physical settings in that state. An inverse function of the extent of dependence on tanker water supply was used as a proxy for sustainability of rural water supply schemes.

Based on scenarios that exist at the district level in the state of Maharashtra with regard to these parameters (effective groundwater recharge potential, aquifer storage space, irrigation water demand, extent of coverage of gravity irrigation, and extent of dependence on wells for water supply), four typologies are identified. Each of the four typologies indicates a certain degree of success of groundwater-based water supply schemes at the district level. These typologies together broadly represent the conditions that can be encountered in the country as a whole in terms of the level of sustainability of groundwater-based rural water supply schemes.

This was followed by an analysis of field-level data to assess the performance of rural water supply schemes on various performance indicators that validated the findings of the modelling studies. Accordingly, various technical strategies for improving the sustainability of rural water supply were suggested for each typology. The most crucial aspect of the development of this technical strategy was the assessment of uncommitted surface water flows in the major river basins that can be utilized for augmenting existing or developing new water supply schemes, based on considerations of available resource base (dependable yield), the water allocated by interstate water disputes tribunals, the demand that exists for the water from the existing schemes, and the topographical conditions that determine the harnessable portion of the surplus resource.

Following the assessment of uncommitted flows that can be tapped in each basin, further analysis explored the ways to use the available surface water in each basin to augment the existing drinking water schemes within those basins and also to transfer to other basins that required additional water to augment the existing poorly-performing schemes or build new ones, using concepts such as the construction of small reservoirs in hills to supply drinking water, 'replacement water', reallocation of surface water from existing reservoir-based schemes from irrigation to domestic use, and 'inter-regional bulk water transfer'. These analyses considered the existing water storage and distribution infrastructures such as reservoirs and canals and the overall water supply–demand situation in different regions and the degree of failure of groundwater-based schemes.

The issues and challenges of and the strategies for managing groundwater quality are then discussed analytically, covering groundwater contamination and pollution problems, the technological and institutional measures for their control, and the concepts and approaches for water quality surveillance in rural areas. Based on the learnings on the key determinants of success and failure of rural water supply schemes in different physical and socio-economic environments, a framework for decision-making that helps plan for better performing rural water supply schemes in India evolved. Applying this framework, and using an assessment of the macro-level situation in the country with regard to the parameters of success of rural water supply performance, specific strategies for improving the sustainability of rural water supply schemes are also proposed in areas where such schemes are likely to perform poorly. Finally, we discuss how this strategy of improving the sustainability of rural water supply in terms of quantity and quality helps the country achieve the goal of functional tap water connection for all rural households, a prerequisite for rural drinking water security, and the cost and economic implications of the same.

1.6 Book Contents and Chapter Outlines

The book has nine chapters. This chapter provides the rationale and scope of the book. It also outlines the remaining six chapters. It gives an overview of the progress made by Indian states and union territories in improving formal water supply in rural areas of the country since Independence. It also discusses the various paradigms and techno-institutional models followed for creating drinking water sources. It also provides an overview of the significant challenges faced by the rural water supply sector from the point of view of sustainability of the resource base, financial sustainability of the schemes and social sustainability.

Chapter 2 presents the analysis of how the hydrological and geo-environmental factors affect groundwater recharge and availability in hard rock areas of India. The analysis uses time series data from Maharashtra, at the district level and the state level. The focus is on the impact of monsoon rainfall, pre-monsoon depth to water levels, recharge potential, and aquifer characteristics. Using the understanding developed on the basis of the statistical analysis performed at the district level, a comprehensive model was developed to explain groundwater behaviour during monsoon across the state that uses parameters such as monsoon rainfall, depth to water levels, recharge potential, and aquifer characteristics and was validated. The chapter explains the development and validation of the model and the practical and policy relevance of the model results.

Chapter 3 presents the investigation into the factors that explain the varying performance of rural water supply schemes. Building on the analysis presented in Chap. 2 that looked at the impact of various factors on groundwater-level fluctuations during monsoon, the analysis in Chap. 3 looks at the potential influence of additional factors that are physical (geomorphology, climate), technical (dependence on groundwater, extent of gravity surface irrigation), and socio-economic (cropped area and irrigated area) in nature.

The chapter describes the statistical model developed and validated to explain the varying performance of water supply schemes under different physical environments using parameters such as monsoon rainfall, pre-monsoon depth to water levels, recharge potential, irrigation demand, extent of dependence of groundwater-based schemes, extent of gravity surface irrigation, geomorphology, aquifer characteristics. An assessment of the likely success of rural water supply schemes in each district was done on the basis of how the districts fared on these parameters. Based on this comparative assessment, four typologies of districts are identified. Each typology required a certain type of intervention to improve the sustainability of water supply schemes. The chapter also explains the practical and policy relevance of the model's results in rural water supply planning.

Chapter 4 presents the results of the analysis of field data on the performance of rural water supply schemes in Maharashtra. A total of 12 schemes and 255 households were surveyed. For the performance review, data sets on water supply scheme characteristics and their operation and maintenance were analysed. These include scheme type, its coverage (villages, habitations, and households), planned and actual supply, access (time taken and distance travelled) and frequency of water supply, maintenance of the system, water tax recovery, and adoption of improved sanitation facilities by the households. The performance assessment covered: physical performance (water supply levels, water supply coverage, duration and frequency of water supply, reliability), economic performance (cost of water supply, capital and O&M), financial performance (rate of recovery of water charges), governance and management issues, institutional performance, etc.

Chapter 5 presents the estimates of un-utilized surface water resources in major river basins of Maharashtra that can be utilized for augmenting existing or developing new water supply schemes. For this purpose, a systematic assessment of water utilization by major and medium and minor reservoir and diversion-based schemes in each basin was made using standard methodologies and the estimates were compared with the total renewable water resources available in those basins. Further, the current proposals for improving the sustainability of rural water supply were analysed and compared against the analysis of water supply schemes' performance (Chap. 4) to determine their usefulness in achieving domestic water security.

Chapter 6 presents a multi-pronged strategy for achieving sustainable rural domestic water supply in Maharashtra. The strategy proposed is based on a thorough understanding of the various physical and socio-economic attributes that influence the performance of groundwater-based schemes and with due consideration to the conditions that exist vis-à-vis these attributes and the availability of unutilized surface water resources in major river basins in hard rock areas. The strategy takes into account the current thinking in the government vis-a-vis the nature of future investments for achieving water security (for all sectors) to make sure that the suggested proposal is aligned. The technical feasibility of implementing various proposed interventions (as per the strategy) along with the estimated unit costs and total investment required are discussed. The cost-effectiveness of the interventions against tanker water supply is also examined.

Moving away from the issue of managing water quantity, in Chap. 7, we discuss groundwater quality issues in the context of rural drinking water security in India. It discusses the extent of contamination and pollution of groundwater in India from the point of view of using it for human consumption and highlights the emerging problem of microbial and nitrate pollution of groundwater due to onsite sanitation through millions of toilets in rural areas. It also discusses the issues and challenges in tackling groundwater contamination and pollution from technical and institutional perspectives, particularly the dispersed microbial and nitrate pollution from rural latrines and non-point pollution from agricultural practices, for protecting it as the source water for domestic supplies. While discussing the technologies to treat poor quality groundwater to reduce the adverse health impacts of using it, it makes a case that preventing microbial and nitrate contamination of groundwater would require surveillance of drinking water sources based on risk assessment.

The absence of a comprehensive and scientific approach to water quality monitoring leads to ineffective and inefficient surveillance of drinking water quality. In Chap. 8, an attempt is made to develop a composite index, which would provide indications as to where the water supply surveillance—in relation to the quality of water—has to be more frequent to avert any public health hazards and where routine monitoring of certain basic parameters would be sufficient. The index helps assess the public health risks associated with the poor quality of water resources in an area. This composite index has three dimensions: Threat, Exposure and Vulnerability. The indices corresponding to these attributes have seven sub-indices in total. The number of 'minor' factors, which together are considered to have an influence on the measure of these sub-indices, the underlying assumptions, the methods and procedure to compute and the data sources are also discussed. The values of the index are computed for all the blocks of Maharashtra, and the variations are explained.

Building on the knowledge gained from the analysis presented in Chaps. 2, 3, 4, 5, 6, 7, and 8, the last chapter (Chap. 9) provides a decision-making framework that helps plan better performing rural water supply schemes. The knowledge is with respect to the attributes that have bearing on the type of water supply schemes that would be sustainable in any rural area, and the manner in which they influence the performance of the schemes. The chapter analyses the macro-level conditions that exist in India vis-à-vis these attributes. The attributes considered are surface hydrology; topography; geology and geohydrology; groundwater chemistry; and nature and degree of over-exploitation of groundwater. Subsequently, the regions in India where groundwater-based schemes are unlikely to be sustainable are identified and the reasons explained. Further, strategies for the provision of rural drinking water supply in those regions from surface water resources are evolved, based on consideration of surface hydrology and topography. The key institutional and policy reforms that are required to manage rural water supply on a sustainable basis are also discussed.

The concluding Chap. 10 discusses the importance of having household tap water connection, for rural water security; illustrates its benefits and impacts; and illustrates, with the support of empirical evidence, the inextricable link between the quality of water supply and the rural people's willingness to invest in household tap connections and how the objective of individual household tap connection in rural areas can be achieved through the provision of treated water from reservoir-based surface water schemes. The chapter also discusses the cost implications of achieving rural drinking water security in India through reservoir-based schemes with piped water supply to households in areas where groundwater-based schemes are unsustainable from the point of view of quality and quantity and the economic benefits from the same.

To conclude, sustainable provisioning of rural drinking water is a widespread and growing concern not only across India, but also globally, from the household to district, state, and national scales (Wescoat and Murthy 2021). With a heavy dependence on groundwater-based schemes for rural water supply in India, the rural drinking water sector faces several sustainability issues ranging from increasing water demand from competing use sectors to depletion of source water (Kumar et al. 2021), premature deterioration of village schemes and services, weak capacity of the scheme operators, frequently changing state and national policies (Wescoat and Murthy 2021), and impacts of climate variability, especially droughts (Kumar et al. 2021). The book offers a framework for assessing and addressing these challenges and also presents strategies for providing sustainable water supply in rural areas of the country, covering different physical and socio-economic conditions.

References

Bassi N, Kabir Y (2016) Sustainability versus local management: comparative performance of rural water supply schemes. In: Kumar MD, Kabir Y, James AJ (eds) Rural water systems for multiple uses and livelihood security. Elsevier, Netherlands, UK and USA, pp 87–115

Bassi N, Kumar MD, Narayanamoorthy A (2014a) Ghost workers and invisible dams: checking the validity of claims about impacts of NREGA. In: Kumar MD, Bassi N, Narayanamoorthy

A, Sivamohan MVK (eds) The water, energy and food security nexus: lessons from India for development. Routledge, UK, pp 61–78

Bassi N, Kumar MD, Niranjan V, Kishan KSR (2014b) The decade of sector reforms of rural water supply in Maharashtra. In: Kumar MD, Bassi N, Narayanamoorthy A, Sivamohan MVK (eds) The water, energy and food security nexus: lessons from India for development. Routledge, UK, pp 194–218

Bassi N, Kabir Y, Ghodke A (2021) Planning of rural water supply systems: role of climatic factors and other considerations. In: Kumar MD, Kabir Y, Hemani R, Bassi N (eds) Management of irrigation and water supply under climatic extremes: empirical analysis and policy lessons from India. Springer Nature, Switzerland, pp 161–177

Baumann E (2006) Do operation and maintenance pay? Waterlines 25(1):10–12

Bolt E, Schouten T, Moriarty P (2001) From systems for service: scaling up community management. In: Scott R (ed) People and systems for water, sanitation and health. 27th WEDC international conference, Lusaka, Zambia, 20–24 Aug 2001

Brunner N, Lele A, Starkl M, Grassini L (2010) Water sector reform policy of India: experiences from case studies in Maharashtra. J Policy Model 32(4):544–561

Cullet P (2011) Realisation of the fundamental right to water in rural areas: implications of the evolving policy framework for drinking water. Econ Polit Wkly 46(12):56–62

FAO (2006) Livestock's long shadow: environmental issues and options. Food and Agriculture Organization of the United Nations, Rome

GoI (Government of India) (2018) Report of the Comptroller and Auditor General of India on performance audit of national rural drinking water programme. Ministry of Drinking Water and Sanitation, Union Government, New Delhi

GoI (Government of India) () Operational guidelines for the implementation of Jal Jeevan Mission (Har Ghar Jal). Department of Drinking Water and Sanitation, Ministry of Jal Shakti, Government of India, New Delhi

Harvey PA, Reed RA (2007) Community-managed water supplies in Africa: sustainable or dispensable? Community Dev J 42(3):365–378

Howard G, Bartram J (2003) Domestic water quantity, service level and health. World Health Organization, Geneva

Howard G, Bartram J (2010) Vision 2030: the resilience of water supply and sanitation in the face of climate change. World Health Organization, Geneva

Hutchings P, Chan MY, Cuadrado L, Ezbakhe F, Mesa B, Tamekawa C, Franceys R (2015) A systematic review of success factors in the community management of rural water supplies over the past 30 years. Water Policy 17(5):963–983

IRAP, CTARA, UNICEF (2018) Compendium of training materials for the capacity building of the faculty and students of engineering colleges on improving the performance of rural water supply and sanitation sector in Maharashtra: under the Unnat Maharashtra Abhiyan (UMA). UNICEF, Mumbai

Jalan J, Ravallion M (2003) Does piped water reduce diarrhea for children in rural India? J Eco 112(1):153–173

Kumar MD, Singh OP (2008) How serious are groundwater over-exploitation problems in India?: a fresh investigation into an old issue. In: Kumar MD (ed) Managing water in the face of growing scarcity, inequity and declining returns: exploring fresh approaches, vol 1. Proceedings of the 7th IWMI-TATA annual partners meet, ICRISAT, Hyderabad, 2–4 Apr 2008

Kumar MD, Ghosh S, Patel A, Singh OP, Ravindranath R (2006) Rainwater harvesting in India: some critical issues for basin planning and research. Land Use Water Resour Res 6:1–17

Kumar MD, Kabir Y, James AJ (eds) (2016) Rural water systems for multiple uses and livelihood security. Elsevier, Netherlands, UK and USA

Kumar MD, Kabir Y, Hemani R, Bassi N (eds) (2021) Management of irrigation and water supply under climatic extremes: empirical analysis and policy lessons from India. Springer Nature, Switzerland

Lockwood H, Smits S (2011) Supporting rural water supply: moving towards a service delivery approach. Practical Action Publishing Ltd., Warwickshire
Moriarty P, Smits S, Butterworth J, Franceys R (2013) Trends in rural water supply: towards a service delivery approach. Water Altern 6(3):329–349
Prokopy LS (2005) The relationship between participation and project outcomes: evidence from rural water supply projects in India. World Dev 33(11):1801–1819
Reddy VR, Rammohan Rao MS, Venkataswamy M (2010) 'Slippage': the bane of drinking water and sanitation sector (a study of extent and causes in rural Andhra Pradesh). WASHCost India-CESS working paper, Hyderabad
Sampat P (2007) Swajaldhara or 'Pay'-jal-dhara: sector reform and the right to drinking water in Rajasthan and Maharashtra. Law Environ Dev J 3(2):3–125
Sangameswaran P (2010) Rural drinking water reforms in Maharashtra: the role of neoliberalism. Econ Polit Wkly 45(4):62–69
Schouten T, Moriarty P (2003) Community water, community management. From system to service in rural areas. ITDG Publishing, London
Smits S, Van Koppen B, Moriarty P, Butterworth J (2010) Multiple-use services as alternative to rural water supply services: a characterisation of the approach. Water Altern 3(1):102–121
UNICEF, WHO (2012) Progress on drinking water and sanitation: 2012 update. UNICEF and WHO, USA
WELL (Water and Environmental Health at London and Loughborough) (1998) Guidance manual on water supply and sanitation programmes. WEDC, Loughborough University, Loughborough
Wescoat JL Jr, Murthy JVR (2021) District drinking water planning for sustainability in Maharashtra: between local and global scales. Sustainability 13(15). https://doi.org/10.3390/su13158288
WSP (Water and Sanitation Programme) (2004) Alternate management approaches for village water supply systems focus on Maharashtra. Water and Sanitation Programme-South Asia, New Delhi

Chapter 2
Factors Influencing Groundwater Behaviour During Monsoon: Analysis from Maharashtra

2.1 Introduction

Groundwater continues to be the main source of water supply in rural areas of India. Its scientific assessment is very crucial for planning groundwater-based schemes, especially the estimation of renewable groundwater resources occurring due to recharge from rainfall during the monsoon. Though a scientific methodology has long been developed for the assessment of groundwater recharge using 'water-level fluctuation approach' that involve estimation of various components of water balance (see Chatterjee and Ray 2014), its application is very limited due to paucity of data, especially on specific yield of the aquifers.

Though data on water level fluctuations during monsoon are generated for a large number of observation wells in each district of the country (see CGWB 2021), the specific yield values that the estimates of groundwater recharge are highly sensitive to, are very difficult to obtain. When the geological formation characteristics vary from location to location, the geohydrological parameters, including specific yield can also vary. Maharashtra is one such state which displays high heterogeneity in formation characteristics, though 90% of the state's geographical area is underlain by hard rock formations with basalt, crystalline rocks and laterite, and the remaining 10% by alluvial deposits.

Water-level trends in wells in hilly and hard rock areas are quite dynamic and complex, and the geo-hydrological setting of the area can induce significant limits on the utilizable recharge from rainfall (Deshpande et al. 2016). Yet, the general perception even among geohydrologists in India is that water level fluctuations in wells during the monsoon is a direct function of the rainfall and that higher water level fluctuations during the monsoon would indicate a higher quantum of recharge. However, often groundwater behaviour during monsoon in the same area in different years is not found to be strongly correlated with the amount of rainfall that occurred in that area in those years (Deshpande et al. 2016, p. 194). Further, higher rise in water levels during monsoon is also not correlated with increased renewable

M. Dinesh Kumar et al., *Drinking Water Security in Rural India*, Water Resources Development and Management, https://doi.org/10.1007/978-981-16-9198-0_2

groundwater, with areas witnessing lower water level fluctuations during monsoon sustaining higher intensity of groundwater use and vice versa.

Often in the absence of data on water level fluctuations and specific yield, groundwater agencies involved in the resource assessment consider a fraction of the rainfall during the monsoon as the recharge by multiplying the rainfall with a coefficient. Such an approach leads to over-estimation of monsoon recharge in many hilly regions that witness very high monsoon rainfall but have low infiltration due to rocky terrains such as the north-eastern hill region and Western Ghats. Secondly, the overemphasis on water level fluctuations during monsoon has led to incorrect evaluation of the performance of artificial recharge schemes, with high water level fluctuations occurring over small pockets around the recharge structures often being attributed to the high effectiveness of the structures. In sum, poor understanding of the complex factors influencing groundwater behaviour during monsoon over space and time leads to unscientific planning of groundwater.

In this study, a statistical model was developed to explain groundwater-level fluctuation during monsoon using data on parameters such as coefficient for recharge potential (which determines the amount of infiltration to the aquifer from a unit of rainfall), annual rainfall and storage space available in the aquifer before the onset of monsoon, and the same was validated. The model development is based on the principle that the maximum water level fluctuations that can occur in any given area during the monsoon is a function of the quantum of recharge (in mm) and the specific yield of the aquifer. Though the quantum of recharge is a function of the rainfall and the 'coefficient for recharge potential' (determined by the infiltration characteristics), it is also limited due to the depth to groundwater level before the onset of monsoon which is a function. The model was developed for the data from the state of Maharashtra, which is characterized by spatial variation in geology, geohydrology, morphology, and spatial and temporal variation in rainfall. For modelling, time series data on groundwater-level fluctuations, rainfall and pre-monsoon depth to water levels, and data on aquifer properties and surface morphology were collected at the district level.

2.2 Review of Literature

The occurrence of groundwater in an area depends on several factors. These include slope, drainage density, land use, geology, lineament density, and geomorphology (Alhassoun 2011; Rajaveni et al. 2017). All these factors influence the infiltration of rainwater into the sub-surface layer. For instance, in the low sloping area, the rainwater would have more time to remain on the ground surface and infiltrate, whereas in the case of highly sloping areas, the run-off is more immediate offering less retention time for the water on the ground surface and thus significantly reducing the groundwater recharge (Rajaveni et al. 2017).

In hard rock areas of India, several studies were undertaken to evaluate the aquifer characteristics such as hydraulic conductivity and transmissivity (Chandra et al. 2008;

Maréchal et al. 2008; Kumar et al. 2016; Machiwal et al. 2017). However, the groundwater behaviour (rise and fall of water level, recharge) in the hard rock formations is perhaps one of the least understood subjects.

The water level fluctuations (WLF) approach, which uses specific yield and average water level fluctuations during the monsoon, is widely used to estimate the groundwater recharge for the shallow aquifers in the hard rock areas (Varalakshmi et al. 2014; Machiwal et al. 2017). The WLF method is based on the premise that rises in groundwater levels in unconfined aquifers are due to recharge water arriving at the water table that immediately goes into storage and assumes that there is no groundwater discharge to streams or springs as baseflow, evapotranspiration from groundwater, and groundwater abstraction through pumping during the recharge period (Healy and Cook 2002).

The popularity of the WLF method can be because of the easy availability of the depth to groundwater-level data and the simplicity of the technique in estimating recharge rates from temporal fluctuations or spatial patterns of groundwater levels (Bhuiyan et al. 2009). Yet, groundwater recharge is one of the most difficult hydrologic parameters to be accurately quantified in the semiarid and arid hard rock regions due to the difficulty in accurately estimating specific yield of the aquifer (Bhuiyan et al. 2009; Machiwal et al. 2017).

Further, in the shallow hard rock aquifers, irrespective of whether groundwater is abstracted or not, it does not remain static and a substantial amount is lost as a base flow after the monsoon rains, and therefore, the recharge estimates only correspond to that during the monsoon and not the utilizable recharge available during the year (Kumar and Singh 2008; Maggirwar and Umrikar 2011; Deshpande et al. 2016). Also, if the rate of recharge is constant or equal to the rate of groundwater drainage, water levels would not change and the WTF method would predict no recharge (Healy and Cook 2002). It was also found that the groundwater levels in such formations are influenced by the topography, presence of structural hills, and the density of pumping wells (Machiwal et al. 2017). The lack of a strong correlation between rainfall and water level fluctuations in the same area, when analysed on a time scale, is also observed (Deshpande et al. 2016).

The literature review clearly shows that even with the same amount of rainfall, there can be significant spatial (Machiwal et al. 2017) and temporal (Deshpande et al. 2016) variation in groundwater recharge in the hard rock aquifers due to an interplay of several complex processes, which eventually would get reflected in water level fluctuations in wells during the monsoon. However, the challenge lies in correctly identifying the relevant variables that influence groundwater recharge during the monsoon and understanding their spatial and temporal variations, so as to capture them in modelling regional groundwater behaviour.

2.3 Methodology

The methodology used for the study is multi-variate analysis involving data on water level fluctuations at the district level as the dependent variable, and the data on average district-level rainfall, average pre-monsoon depth to water level at the district level, and specific yield of the aquifer and recharge coefficient as independent variables. The approach used is 'inverse parametric modelling', by which the hydrological and geohydrological parameters used in the model (recharge coefficient and the specific yield values) were adjusted for the model calibration so as to get a strong regression coefficient for the statistical relationship with the water level fluctuations.

The model development involved a multi-stage analysis. In the first stage, the effect of the magnitude of rainfall on groundwater-level fluctuation during monsoon was examined by comparing spatial and temporal values of district-level rainfall on spatial and temporal values of average water level fluctuations, respectively, and the same was established. Subsequently, the effect of another parameter, i.e. 'pre-monsoon depth to water levels' along with that of rainfall on water level fluctuations, was examined by performing a regression analysis with time series data of 'average rainfall' and average 'pre-monsoon depth to water levels' as independent variables against 'average water level fluctuations during monsoon' as the dependent variable, in two districts of Maharashtra with no variation in geohydrological properties.

In the third stage, a multi-variate analysis was performed with data from all the districts of Maharashtra, with variations in geohydrology, rainfall, and soil infiltration characteristics to analyse the effect of rainfall, pre-monsoon depth to water levels, and specific yield of the aquifer, 'recharge per unit area (multiple of the 'coefficient for recharge potential' and rainfall). The 'coefficient for recharge potential' considered the different landforms of Maharashtra and the map showing priority areas for artificial recharge, prepared by GSDA, Maharashtra. The value of the coefficient chosen was highest for areas with alluvial valleys and plains that provide favourable conditions for rainwater to infiltrate and lowest for steep and hard rock terrains that cause rainwater to run off without infiltrating.

2.4 Rainfall: Seasonality and Other Features

Maharashtra state has 36 districts that are grouped in six administrative divisions, namely Amravati, Aurangabad, Konkan, Nagpur, Nashik, and Pune (refer to Fig. 2.1). However, India Meteorological Department (IMD) has divided Maharashtra state into four meteorological subdivisions. These are (1) Konkan, receiving very high rainfall during monsoon season; (2) the Vidarbha region that receives less annual rainfall than that of Konkan but more than the other two subdivisions; (3) Central Maharashtra, receiving more annual rainfall than the one subdivision; and (4) the Marathwada, which receives the lowest rainfall in the state. The mean average annual rainfall (1901–2017) is 2988 mm for Konkan, 1094 mm for Vidarbha, 881 mm for Central

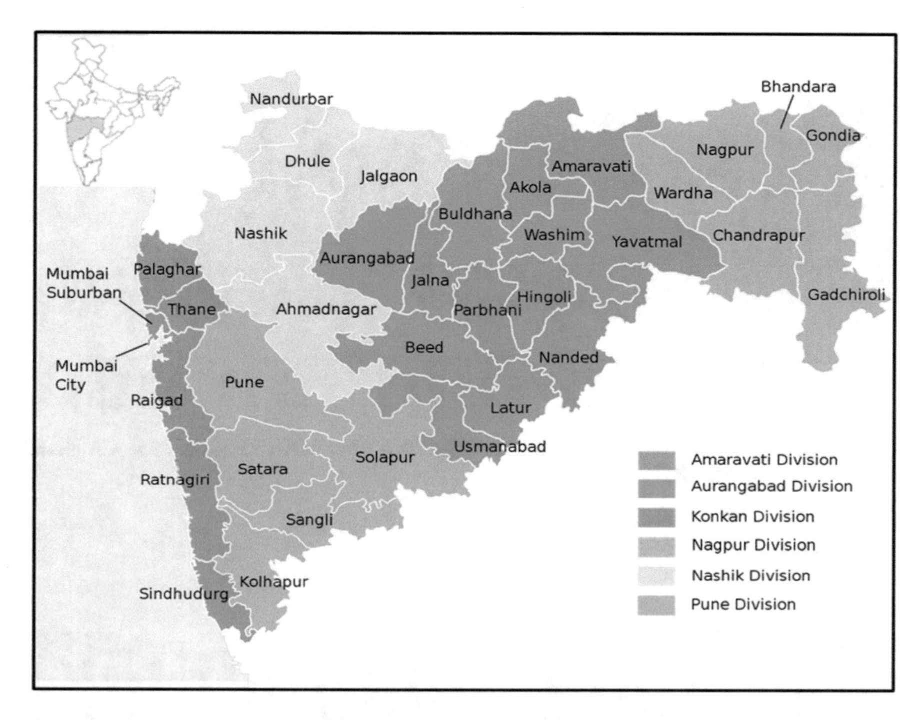

Fig. 2.1 Districts and administrative divisions of Maharashtra. *Source* Wikipedia under Creative Commons License

Maharashtra, and 792 mm for Marathwada. In all the divisions, rainfall exhibits high inter-annual variability. Further, the spatial variability in rainfall with respect to divisions is also remarkable (Fig. 2.2).

The distribution of rainfall over the season is as important as the total annual amount of rainfall while evaluating its impact on hydrology, ecology, agriculture, and water use as they greatly influence the partitioning of water into run-off, evapotranspiration, soil infiltration, and groundwater recharge (Xiao and Moody 2004; Small et al. 2006; Guhathakurta and Saji 2013). To capture the extent of variability in rainfall during the monsoon season, Guhathakurta and Saji (2013) computed the coefficient of variation (CV) in rainfall in different districts of Maharashtra based on data for the period from 1901 to 2006. The CV value for all the districts in Maharashtra is presented in Fig. 2.3. It was found to be varying from a lowest of 16.8 in Sindhudurg and second lowest of 17.9 in Ratnagiri to a highest of 57.0 in Satara and second highest of 35.2 in Mumbai sub-urban district.

Further, Guhathakurta and Saji (2013) computed the long-term trend in the monthly and seasonal rainfalls. Some of the important results which have implications for hydrological planning and water resources development and management were:

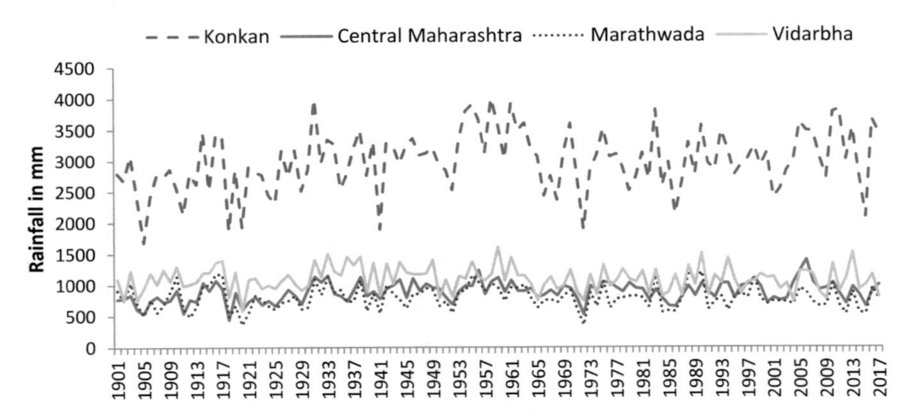

Fig. 2.2 Annual spatial average of rainfall for different regions of Maharashtra. *Source* Authors' analysis based on IMD data

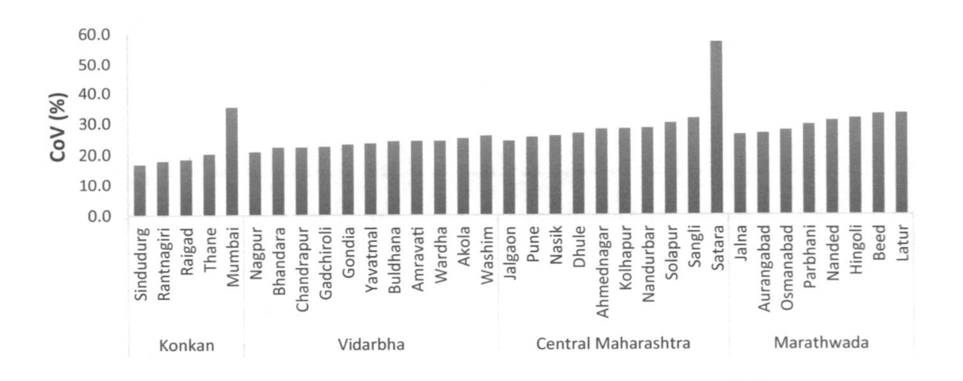

Fig. 2.3 Extent of variability in monsoon rainfall, Maharashtra. *Source* Authors' analysis based on data presented in Guhathakurta and Saji (2013)

(1) Across the four seasons, the zone corresponding to the mean maximum rainfall shifts from the eastern parts of the state in winter to southern parts in pre-monsoon and north-western parts in monsoon and post-monsoon seasons.

(2) As is the case with most of the other states in India, the majority of the rainfall occurs during the monsoon season in Maharashtra. The variability in rainfall during monsoon and post-monsoon is substantially lower than that of the winter and the summer seasons.

(3) During the winter and the pre-monsoon months, many districts exhibited a decreasing trend in rainfall. In the north and central parts, this trend is visible in January, in the north-eastern parts in February, and in the coastal districts in May.

(4) The quantum of rainfall during monsoon (mainly in August and October) has increased significantly. For most of the districts in Central Maharashtra and the Konkan region, there is a significant increase in rainfall. For August, the average increase in rainfall varies from 0.57 to 2.84 mm/year (at 95% level of significance). However, some districts in Marathwada and Vidarbha have shown decreasing trend in the rainfall trend. For the month of July, the average decrease in rainfall varies from 0.77 to 1.39 mm/year (at 95% level of significance).

(5) There was no significant trend in the monthly rainfall in any of the districts for November and December.

Thus, over the 100-year period (1901–2006), rainfall has increased significantly in the districts of the Konkan belt and southern and western parts of Maharashtra. Most of the rainfall is received within 2–3 months during the monsoon. However, in Marathwada and Vidarbha regions, there is a significant decrease in the quantum of the rainfall, but with better distribution across the monsoon months.

2.5 Impact of Hydrological and Geohydrological Factors on Groundwater Behaviour

2.5.1 Rainfall and Water Level Fluctuations During Monsoon

Conventional wisdom suggests that rainfall influences natural recharge to groundwater, especially in regions with shallow aquifers. Further, the response of water level to rainfall is influenced by the soil type. For instance, the red soils allow quick infiltration and percolation of rainwater thus resulting in better recharge, whereas the recharge rate for the wells located in the areas with black soils is slow (Arya et al. 2018).

Nevertheless, the general consideration is that the higher the rainfall, the higher the quantum of recharge, a key indicator of which is the water level fluctuations in the monsoon season (Dhar et al. 2014; Bharathkumar and Mohammed-Aslam 2018; Chen et al. 2021), during which most of the rainfall occurs in India. Because of this reason, even in the estimation of renewable groundwater, recharge is often estimated as a percentage of the total annual rainfall in an area, in the absence of reliable data on water table fluctuations, an important variable used in the estimation of renewable groundwater recharge during the monsoon season in the current groundwater estimation methodologies (refer to CGWB 2021).

2.5.2 *Rainfall, Depth to Water Levels, and Water Level Fluctuations*

In order to understand the effect of rainfall on groundwater conditions, various types of analysis were performed. The analysis took into account the fact that rainfall varies from year to year in the same region/district (refer Fig. 2.4) and also varies from region to region and district to district in the same year. Hence, taking the mean values for different years would help reduce the anomalies that might occur when we consider only a single year, due to extreme conditions such as droughts and abnormal wetness that an area can experience in a given year. In the first analysis, mean values of the spatial average rainfall of different districts for the time period of 1999–2018 were taken and compared against the mean values of the spatial average of 'depth to groundwater level' in each district for the same period, i.e. 1999–2018. The values of the two variables are presented in Table 2.1 and its graphical representation in Fig. 2.5. Regression analysis shows a strong correlation between the two variables, with an R-square value of 0.56. The relationship was found to be exponential. Higher the rainfall, lower the depth to water table (Fig. 2.6).

In the next round of analysis, the mean value of depth to water level (for all observation wells in Maharashtra) for different years was compared against the spatial

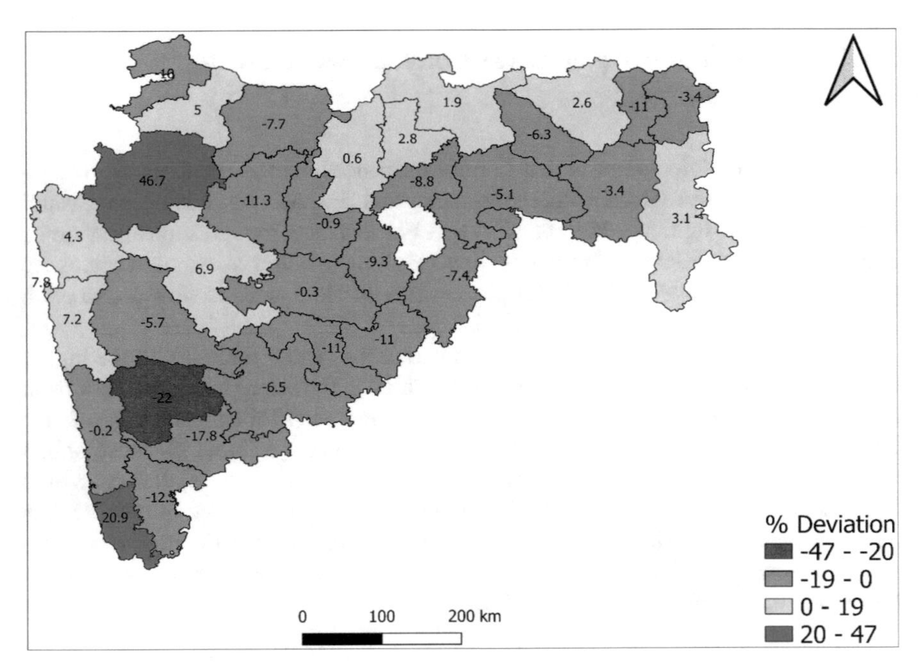

Fig. 2.4 Average deviation in annual rainfall from normal (1998–2019) within districts of Maharashtra

Table 2.1 Mean of spatial average rainfall (mm) and depth to water level (m) across the districts of Maharashtra (1998–2019)

District	Rainfall (mm)			Depth to water level (m)	
	Actual (mm)	Normal (mm)	Deviation (%) from normal	Water level (m)	No. of stations
Ahmadnagar	636.8	595.7	6.89	9.25	100
Akola	848.4	825.3	2.80	12.53	50
Amravati	904.0	886.4	1.99	9.79	156
Aurangabad	675.1	760.9	−11.27	9.93	62
Bhandara	1211.3	1361.3	−11.02	8.33	41
Bid	732.4	734.4	−0.27	8.63	87
Buldana	797.1	792.5	0.58	9.86	90
Chandrapur	1261.3	1305.4	−3.38	7.25	90
Dhule	679.4	647.0	5.01	9.26	51
Garhchiroli	1473.2	1428.5	3.13	7.38	73
Gondiya	1331.6	1377.9	−3.36	7.51	41
Hingoli	868.6	–	–	9.36	46
Jalgaon	722.3	782.4	−7.68	13.03	88
Jalna	745.5	752.2	−0.89	9.52	75
Kolhapur	1426.1	1625.4	−12.26	5.79	53
Latur	826.9	929.4	−11.03	11.53	56
Mumbai	2414.8	2252.7	7.19	3.65	7
Mumbai suburban	2725.5	–	–	4.56	18
Nagpur	1109.7	1082.1	2.55	7.83	117
Nanded	906.2	978.5	−7.38	8.52	87
Nandurbar	822.3	914.4	−10.07	11.43	37
Nashik	965.4	658.0	46.71	8.39	108
Osmanabad	751.6	844.6	−11.01	9.50	69
Parbhani	830.7	915.9	−9.31	9.68	59
Pune	1058.3	1122.6	−5.73	6.05	87
Raigarh	3558.1	3320.2	7.16	3.76	53
Ratnagiri	3284.2	3290.4	−0.19	7.94	81
Sangli	606.1	737.2	−17.79	7.65	64
Satara	871.6	1117.6	−22.01	6.90	75
Sindhudurg	3490.5	2887.2	20.90	6.87	57
Solapur	616.2	659.0	−6.50	8.75	83
Thane	2632.7	2525.2	4.26	4.55	80
Wardha	977.7	1043.2	−6.28	7.82	84
Washim	880.2	965.3	−8.81	8.66	55
Yavatmal	979.5	1032.1	−5.10	7.75	119

Source Prepared using data from IMD and CGWB

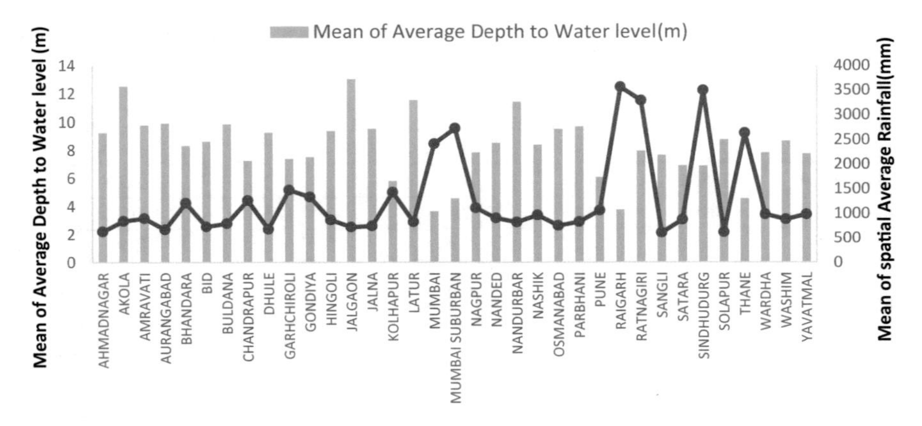

Fig. 2.5 Mean of average depth to water level and spatial average rainfall across the districts of Maharashtra (1998–2019). *Source* Authors' analysis using data from IMD and CGWB

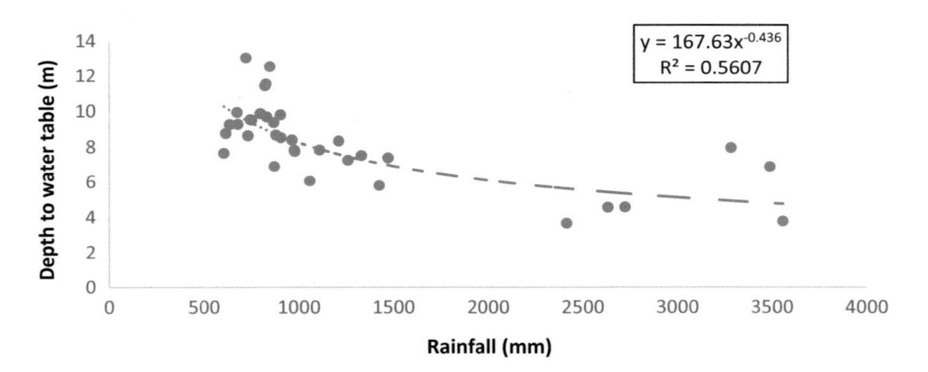

Fig. 2.6 Mean of spatial average depth to water level versus mean of spatial average rainfall in Maharashtra districts. *Source* Authors' analysis using data from IMD and CGWB

average of the annual rainfall of respective years in the state. The graphical representation of outputs is presented in Fig. 2.7. Univariate analysis was performed to see whether any relationship exists between the average annual rainfall and the mean value of depth to water table. The analysis showed a good correlation between the two (R-square = 0.45). Higher rainfall meant a lower value of (mean) depth to water table in the state (see Fig. 2.8). The spatial average rainfall in the state varied from a lowest value of 880 mm in 2015 to a maximum value of 1485 mm in 2019. The mean depth to water table was lowest in 2019.

Further analysis was carried out to see the difference in the impact of rainfall on groundwater conditions between a drought year and a wet year when the rainfall situation changes remarkably. For this, analysis was carried out for two divisions of Maharashtra, viz. Nashik (in Central Maharashtra) and Konkan. For the analysis of Nashik data, the year 2012 was taken as the dry year and 2006 was the wet year. The

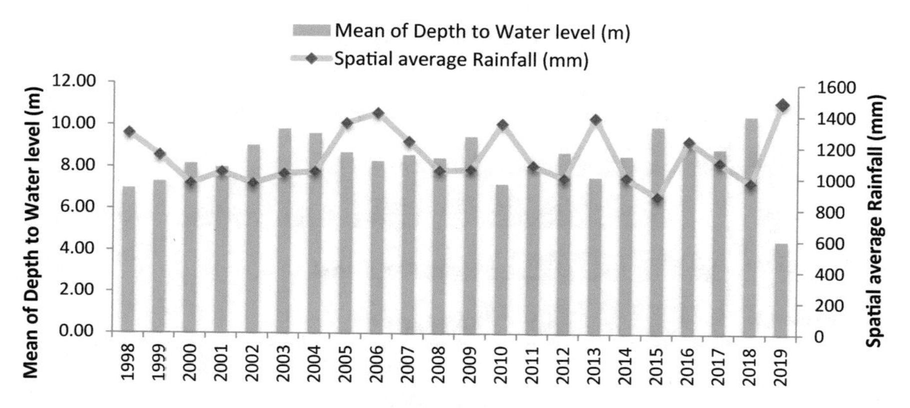

Fig. 2.7 Mean depth to water level and spatial average rainfall in Maharashtra (1998–2019). *Source* Authors' analysis using data from IMD and CGWB

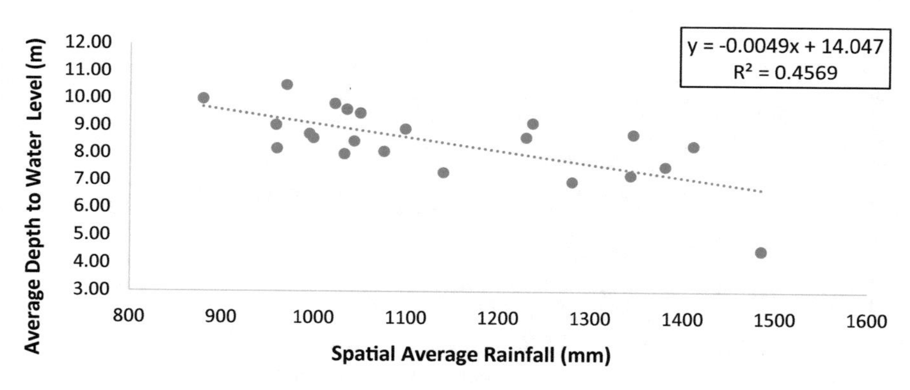

Fig. 2.8 Average depth to water level versus average rainfall in Maharashtra. *Source* Authors' analysis using data from IMD and CGWB

first round of analysis considered data pertaining to different districts. The results are presented in Figs. 2.9 and 2.10, for the wet year and dry year, respectively. It shows that in all the five districts in the Nashik division, viz. Ahmadnagar, Dhule, Jalgaon, Nashik, and Nandurbar, the water level fluctuations during the monsoon was much higher during the wet year, when compared to the dry year. The water level fluctuations was to the tune of 3.87 m, 4.79 m, 7.24 m, 4.61 m, and 4.67 m for Ahmadnagar, Dhule, Jalgaon, Nashik, and Nandurbar, respectively, during the wet year. Against this, during the dry year, the corresponding fluctuation was to the tune of 0.39 m, 0.40 m, 1.42 m, 2.30 m, and 1.51 m, respectively. Hence, the water level rise during the dry year is very low. In addition to poor recharge, one reason could be that during the drought year farmers might pump water to irrigate their crops due to water stress, bringing down the water table further down. For the division as a

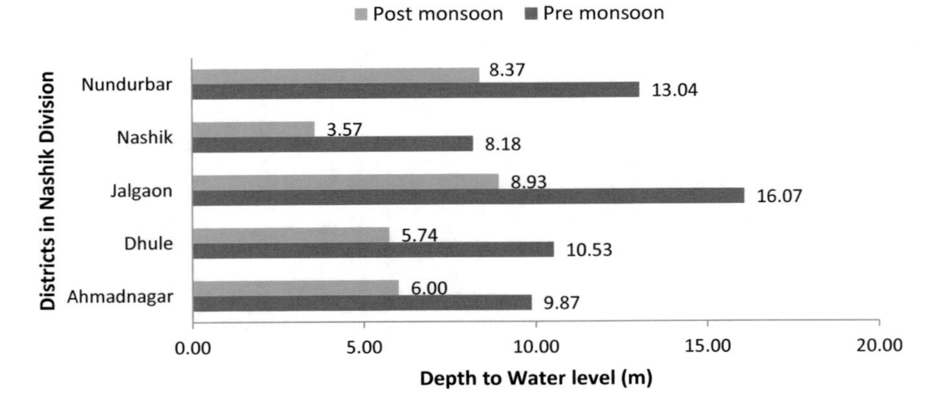

Fig. 2.9 Water level fluctuations in districts of Nashik division during wet year (2006). *Source* Authors' analysis using CGWB data

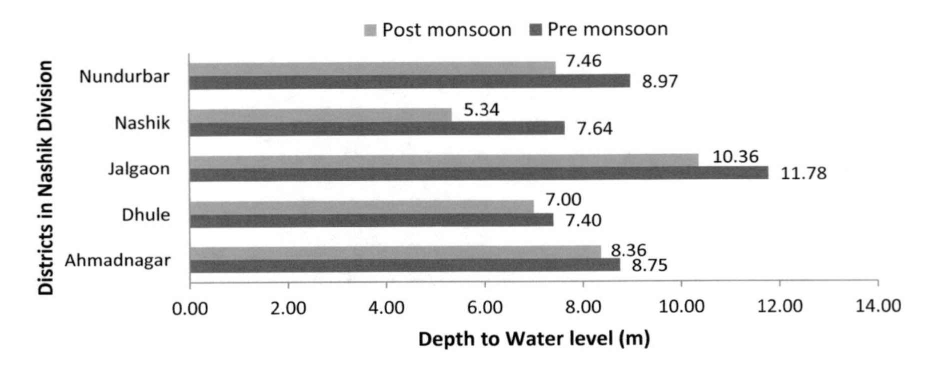

Fig. 2.10 Water level fluctuations in districts of Nashik division during dry year (2012). *Source* Authors' analysis using CGWB data

whole, the water level fluctuations during the wet year was estimated to be 4.60 m, while that of drought year was 1.21 m.

Similar analysis was carried out for the districts in the Konkan division, viz. Mumbai, Mumbai Suburban, Raigarh, Ratnagiri, Sindhudurg, and Thane. For the Konkan division, the year 2015 was taken as the dry year and 2019 was the wet year. The results are presented in graphical form in Figs. 2.11 and 2.12, for the wet year and dry year, respectively. The comparison showed more or less the same pattern. The water level fluctuations during the wet year was 2.8 m, 2.6 m, 2.8 m, 2.3 m, 3.7 m, and 4.7 m, respectively, for Mumbai, Mumbai suburb, Raigarh, Ratnagiri, Sindhudurg, and Thane. The corresponding values for the drought year were 1.6 m, 4.0 m, 2.0 m, 2.2 m, 2.7 m, and 2.2 m, respectively. Therefore, except for one district (i.e. Mumbai suburb), the pre-post-monsoon water level fluctuations during the wet year was higher than that of the drought year. Overall, for the whole of the division,

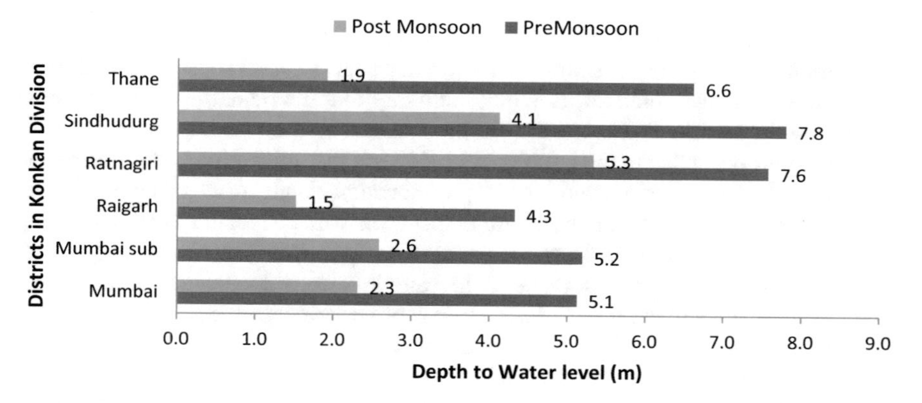

Fig. 2.11 Water level fluctuations in different districts of Konkan division during wet year (2019). *Source* Authors' analysis using CGWB data

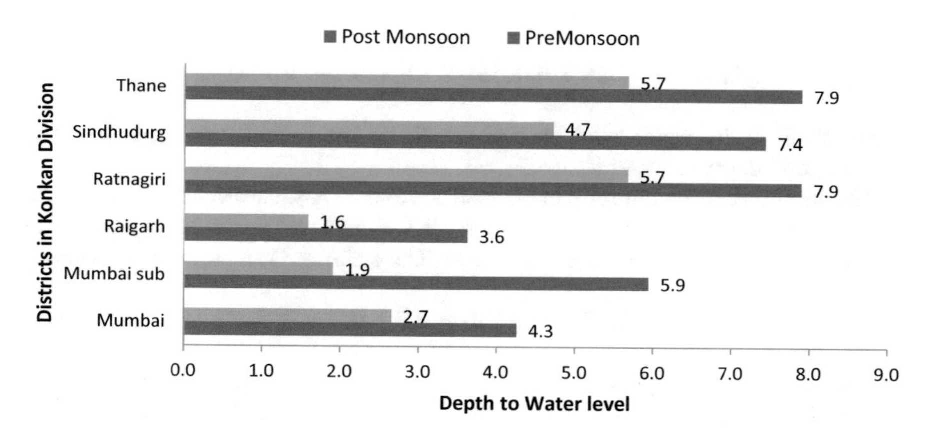

Fig. 2.12 Water level fluctuations in different districts of Konkan division during dry year (2015). *Source* Authors' analysis using CGWB data

the average water level fluctuations was 3.1 m during the wet year, against 2.6 m during the dry year.

2.5.3 *Influence of Recharge Potential and Aquifer Properties on Water Level Fluctuations*

The foregoing analyses suggest that rainfall is a very important factor explaining the water level fluctuations and post-monsoon depth to water level. However, it is also clear that rainfall is not the only factor. Had it been so, the relationship should have been much stronger (refer to Figs. 2.6 and 2.8). Given the fact that around 90% of the geographical area of Maharashtra is underlain by shallow, hard rock formation, with very limited storage potential, it is quite unlikely that the recharge would increase linearly with rainfall. Storage space in the aquifer could be an important determinant. The storage space in the aquifer, for receiving the infiltrating water during the monsoon, is heavily influenced by the pre-monsoon depth to water table. Hence, to examine the influence of the storage space on water level fluctuations during monsoon, an indicator of recharge, regression analysis was carried out with spatial average rainfall and pre-monsoon depth to water level as independent variables, and average water level fluctuations (during monsoon) as the dependent variable for two districts in Maharashtra, viz. Ahmednagar and Aurangabad.

The analysis for Ahmednagar showed that rainfall and pre-monsoon depth to water level explained water level fluctuations to an extent of 66.8% in the case of Ahmadnagar district. The influence of the two variables was significant, at 1% level for pre-monsoon water level and 7% for rainfall. The estimated regression equation is:

$$\mathrm{WLF} = -8.86 + 0.00551 \times \mathrm{RF} + 0.939 \times \mathrm{PrM_{Dwl}}$$

where WLF is the water level fluctuations RF is the rainfall and $\mathrm{PrM_{Dwl}}$ is the pre-monsoon depth to water level.

The analysis carried out for Aurangabad showed an *R*-square value of 0.738. The influence of the two variables in changing the water level fluctuations during the monsoon season was significant at 1% level. The estimated regression equation is:

$$\mathrm{WLF} = -7.50 + 0.00720 \times \mathrm{RF} + 0.669 \times \mathrm{PrM_{Dwl}}$$

These results suggest that water table conditions that exist prior to monsoon are also important factors influencing the total recharge during the monsoon. These results, however, need to be interpreted very carefully though. It does not mean that if we dewater the aquifer before the arrival of monsoon it would result in greater recharge. Instead, what it means is that the real benefits of improved recharge that can result from higher rainfall (in abnormally wet years) can be obtained if and only if there is sufficient storage space available. Non-availability of sufficient storage space to receive the infiltrating water would result in a proportion of the water getting converted into run-off. It must also be noted here that the actual quantum of recharge during the monsoon is not merely a function of water level fluctuations alone. It also depends on the characteristics of the aquifer, i.e. the specific yield or effective porosity. If the specific yield of the aquifer is high (like in an alluvial aquifer), then the water level fluctuations resulting from a given amount of recharge will be lower as compared to that in a situation when the specific yield of the aquifer is very low (like in basalt rock formations that underlie 90% of Maharashtra), with the same amount of recharge. The geological map of Maharashtra is presented in Fig. 2.13.

Further analysis was carried out using the water level fluctuations data for all the districts of Maharashtra. Here again, the mean value of the spatial average of rainfall,

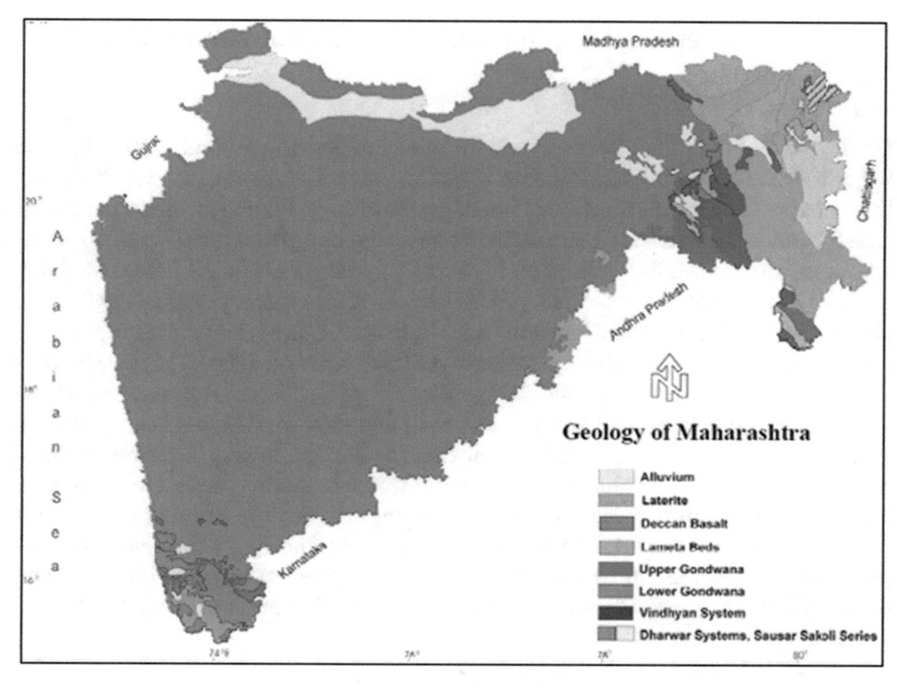

Fig. 2.13 Geological profile of Maharashtra. *Source* Groundwater Surveys and Development Agency (GSDA), Government of Maharashtra

pre-monsoon depth to water level, and WLF was used for the regression. However, the analysis showed that only pre-monsoon depth to water level had an influence on WLF. Rainfall was found to have no effect on WLF. This could be due to the reason that the features that affect the recharge from rainfall (i.e. infiltration characteristics of the ground surface) change significantly across the state—with the most favourable conditions in the alluvial valley (in Tapi basin), to the most unfavourable conditions in the rocky terrains of western Maharashtra (Sangli, Solapur, Satara, Pune, and Ahmadnagar) and eastern Maharashtra (Chandrapur, Nagpur, Gondia, Gadchiroli, and Bhandara). However, this factor is never considered in estimating the recharge potential in the past, with the result that the renewable monsoon recharge gets over-estimated for districts with high rainfall, despite them having unfavourable terrain conditions (Kumar and Singh 2008).

2.6 The Model Explaining Groundwater Behaviour During Monsoon

On the basis of the foregoing analyses, the model explaining groundwater behaviour in hard rock areas during monsoon was developed considering the features that affect the recharge from rainfall and the effect of geohydrology. For the former, a parameter called 'recharge coefficient' was used to estimate recharge from rainfall, and recharge was computed as rainfall multiplied by recharge coefficient. In the process, rainfall as a variable is kept out of the analysis. Selection of values for 'recharge coefficient' considered the different landforms of Maharashtra (Fig. 2.14) and recharge potential map or the map showing priority areas for artificial recharge (Fig. 2.15) prepared by the Groundwater Survey and Development Agency (GSDA), Government of Maharashtra. The value of recharge coefficient used was 0.08 for districts having alluvial valley portions (Akola, Amravati, Buldhana, Dhule, and Jalgaon), which had high recharge potential. The values used for districts with basalt but located in the plateau was 0.05. For districts located in the Konkan region with slightly more favourable conditions for recharge (viz. Sindhudurg, Raigad, and Ratnagiri), the value assumed was 0.07. For districts in eastern Maharashtra, comprising Nagpur, Gondia, Chandrapur, Gadchiroli, and Bhandara with highly undulating and hard rock terrain, the

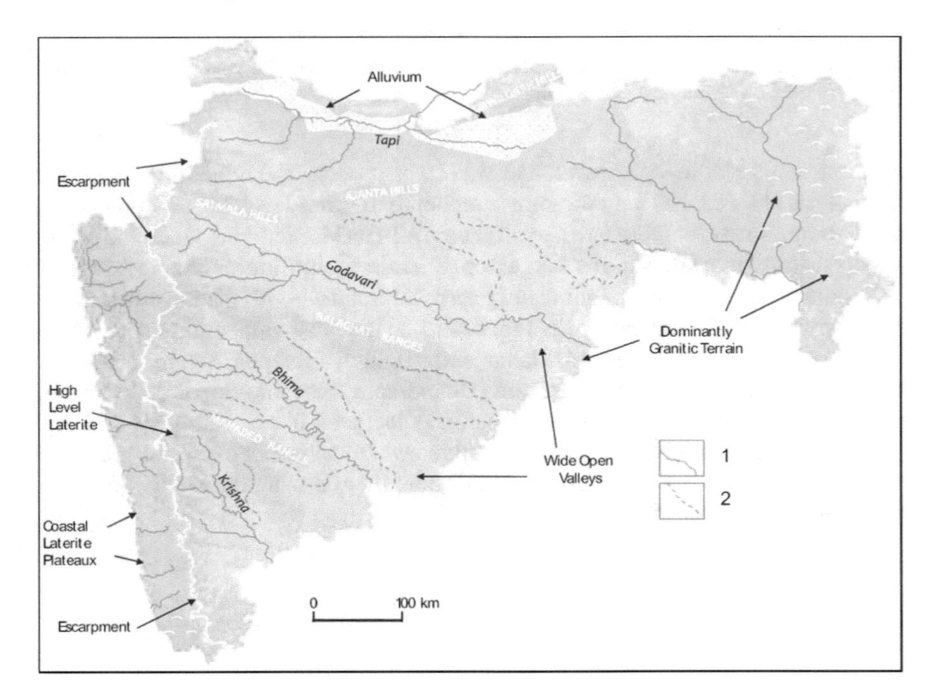

Fig. 2.14 Landforms in Maharashtra. *Source* GSDA, Government of Maharashtra

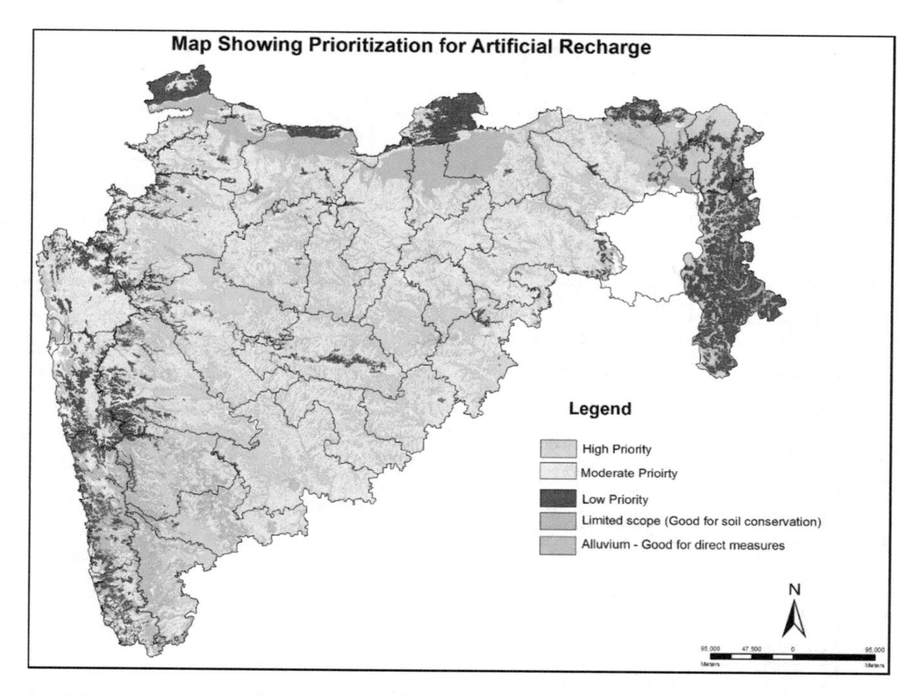

Fig. 2.15 Groundwater recharge potential (priority areas for artificial recharge) of Maharashtra. *Source* GSDA, Government of Maharashtra

value used was 0.03. For western Maharashtra with the most unfavourable conditions for recharge due to steep terrain conditions for most parts, the value used was 0.02.

In order to capture the effect of geohydrology in determining the water level fluctuations, the recharge values were divided by the specific yield values assumed to arrive at a function called WLF coefficient. Selection of specific yield values considered the geological map of Maharashtra (refer to Fig. 2.13). The specific yield values assumed was 0.10 for the districts having alluvial deposits, viz. Akola, Amravati, Buldhana, Dhule, and Jalgaon. The value was 0.02 for all districts having basalt formations. The value used was 0.015 for the districts in eastern Maharashtra, which had crystalline formations (viz. Nagpur, Gondia, Chandrapur, Gadchiroli, and Buldhana). For the Konkan districts with alluvium and laterite along with basalt (viz. Raigad, Ratnagiri, and Sindhudurg), a value of 0.08 was used. The mean values of spatial average rainfall, the assumed values of recharge coefficient, the estimated recharge rate, the assumed values of specific yield, and the estimated values of 'water level fluctuations function' for the rural districts of Maharashtra are presented in Table 2.2. The district-wise specific yield, estimated recharge rate, and WLF function values are also presented in Figs. 2.16, 2.17, and 2.18, respectively.

Regression analysis with water level fluctuations function (Rainfall × Recharge Coefficient/Specific Yield) and the pre-monsoon depth to water level against average WLF showed an *R*-square value of 0.59. Both the parameters had a positive effect

Table 2.2 Mean annual rainfall, recharge coefficient, specific yield, recharge rate, and WLF function for different rural districts of Maharashtra

Sr. No.	District	Spatial average rainfall (mm)	Coeff. for recharge potential	Specific yield of the aquifer	Recharge per unit area (mm) = rainfall × coeff. for recharge potential	Water level fluctuations function = recharge per unit area/sp. yield
1	Ahmednagar	636.8	0.035	0.020	22.28	1114.00
2	Akola	848.4	0.080	0.100	67.872	678.72
3	Amravati	904	0.080	0.100	72.320	723.20
4	Aurangabad	675.1	0.050	0.020	33.755	1687.75
5	Beed	732.4	0.050	0.020	36.620	1831.00
6	Bhandara	1211.3	0.030	0.015	36.339	2422.60
7	Buldhana	797.1	0.080	0.100	63.768	637.68
8	Chandrapur	1261.3	0.030	0.015	37.839	2522.60
9	Dhule	679.4	0.080	0.100	54.352	543.52
10	Gadchiroli	1473.2	0.030	0.015	44.196	2946.40
11	Gondia	1331.6	0.030	0.015	39.948	2663.20
12	Hingoli	868.6	0.050	0.020	43.430	2171.50
13	Jalgaon	722.3	0.080	0.100	57.784	577.84
14	Jalna	745.5	0.050	0.020	37.275	1863.75
15	Kolhapur	1426.1	0.020	0.020	28.522	1426.10
16	Latur	826.9	0.050	0.020	41.345	2067.25
17	Nagpur	1109.7	0.030	0.015	33.291	2219.40
18	Nanded	906.2	0.050	0.020	45.310	2265.50
19	Nandurbar	822.3	0.050	0.020	41.115	2055.75
20	Nashik	965.4	0.050	0.020	48.270	2413.50
21	Osmanabad	751.6	0.050	0.020	37.580	1879.00
22	Palghar	2632.7	0.050	0.020	131.635	6581.75
23	Parbhani	830.7	0.050	0.020	41.535	2076.75
24	Pune	1058.3	0.035	0.020	37.03	1851.50
25	Raigad	3558.1	0.070	0.080	249.067	3113.34
26	Ratnagiri	3284.2	0.070	0.080	229.894	2873.68
27	Sangli	606.1	0.020	0.020	12.122	606.10
28	Satara	871.6	0.020	0.020	17.432	871.60
29	Sindhudurg	3490.5	0.070	0.080	244.335	3054.19
30	Solapur	616.2	0.020	0.020	12.324	616.20
31	Thane	2632.7	0.070	0.080	184.29	2303.61

(continued)

Table 2.2 (continued)

Sr. No.	District	Spatial average rainfall (mm)	Coeff. for recharge potential	Specific yield of the aquifer	Recharge per unit area (mm) = rainfall × coeff. for recharge potential	Water level fluctuations function = recharge per unit area/sp. yield
32	Wardha	977.7	0.050	0.020	48.885	2444.25
33	Washim	880.2	0.050	0.020	44.010	2200.50
34	Yavatmal	979.5	0.050	0.020	48.975	2448.75

Source Authors' own analysis

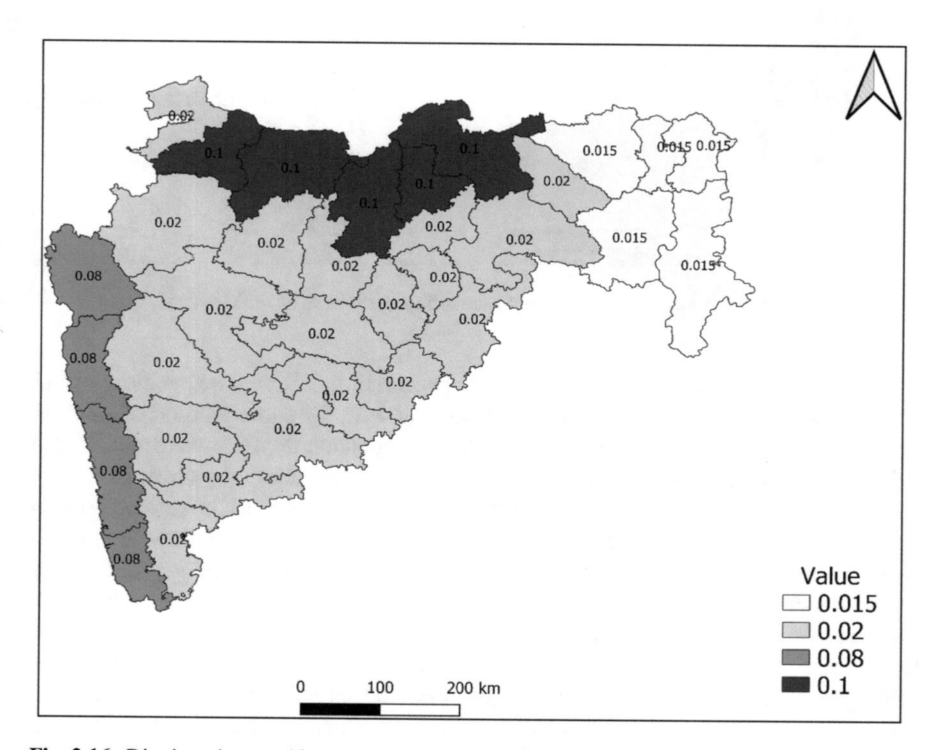

Fig. 2.16 District-wise specific yield of aquifers in Maharashtra

on the WLF values. The estimated regression equation is:

$$\mathrm{WLF} = -0.15 + 0.000372\left(\mathrm{RF} \times \mathrm{Rech_C}/S_y\right) + 0.351\,\mathrm{PrM_{Dwl}}$$

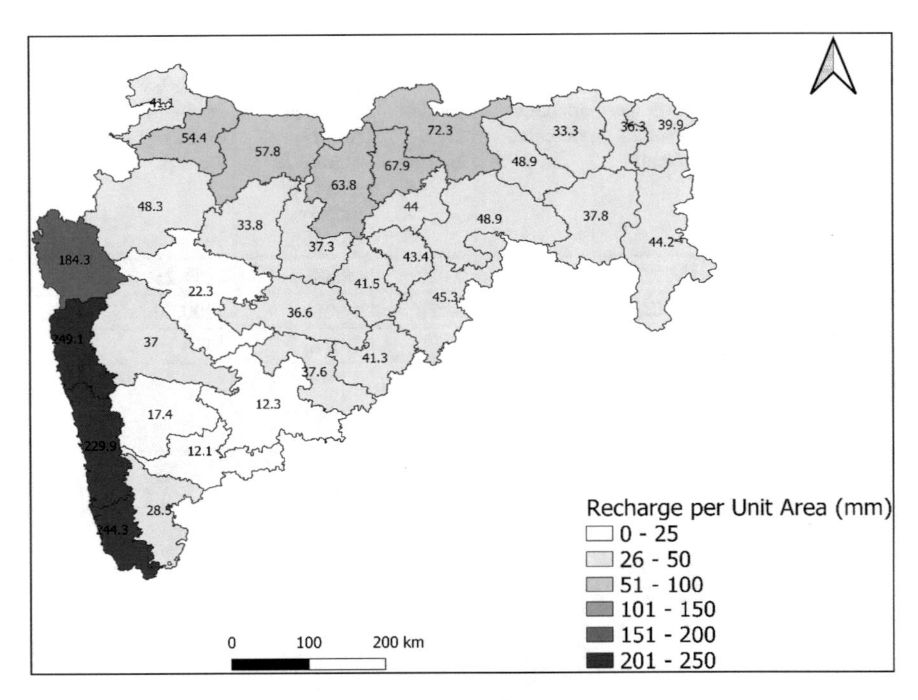

Fig. 2.17 Estimated groundwater recharge rate in different districts of Maharashtra

where $Rech_C$ is the coefficient of recharge potential and S_y is the specific yield of the aquifer. The equation basically suggests that with the same amount of rainfall, and with the same situation with respect to pre-monsoon depth to water levels and specific yield of the aquifer, higher value of recharge coefficient (by virtue of having favourable topsoil conditions for infiltration of the rainwater into the ground) would mean higher WLF as the quantum of recharge would be more. At the same time, with the same rainfall, and low recharge coefficient resulting in poor groundwater recharge, the WLF can be high, if the specific yield value is very poor for the formation. Hence, simple consideration of WLF cannot be the guiding factor for recharge to the aquifer. A good monsoon would indicate the most favourable situation for groundwater recharge in a region with good aquifer characteristics if it is followed by a significant rise in water levels. The model suggests that for Maharashtra state if geomorphological conditions are favourable, a high pre-monsoon depth to water level and a good rainfall would increase chances of an area (here, a district) getting considerable groundwater recharge. It should be mentioned here that there is some amount of arbitrariness in the selection of values for recharge coefficient and specific yield values, which is based on a broader understanding of the geomorphological conditions of the state and characteristics of the aquifers.

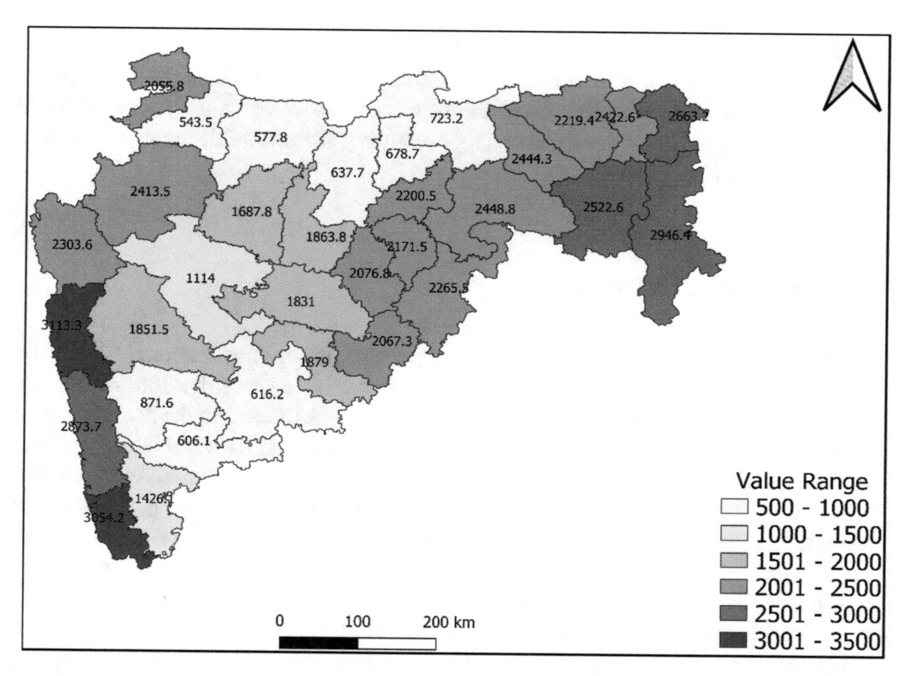

Fig. 2.18 Estimated value of groundwater level fluctuation function for different districts of Maharashtra

2.7 Findings and Conclusion

The statistical model developed to explain groundwater behaviour during monsoon shows that the water level fluctuations during the season is not merely a function of rainfall but the terrain conditions that determine the infiltration, the specific yield of the aquifer, and the pre-monsoon depth to water table. The model can be interpreted in the following ways: having very high rainfall and high infiltration does not guarantee high recharge to the aquifer. There should be sufficient storage space in the aquifer prior to the onset of the monsoon. The regression analysis does suggest that in many situations the benefit of increased recharge from higher rainfall is not realized in an area due to the low storage space available. Even if an area receives excessively high rainfall, if the aquifer has a low specific yield, the terrain is rocky and hilly, and the pre-monsoon depth to water table is high, it can create the most unfavourable conditions for groundwater recharge occurrence. On the other hand, if the area has plain topography with alluvial soils that create favourable conditions for infiltration of rainwater, and if the aquifer has a high specific yield, good rainfall can ensure a large amount of recharge, even if the water table before the onset of monsoon remained high because the rise in water table that can be induced by the infiltrating water will not be very high to cause rejected recharge.

On the other hand, in hard rock areas with very poor yield characteristics of the formation, even a low magnitude of natural recharge from rainfall can cause a significant rise in water table after the monsoon, whereas in an unconsolidated formation, even a large amount of recharge resulting from rainfall may not result in high water table fluctuation.

2.8 Policy Inferences

The findings of the modelling study have important policy implications. First: they suggest that estimating monsoon recharge merely on the basis of the rainfall without factoring in the infiltration characteristics of the ground surface and the actual storage space in the aquifer (a multiple of the pre-monsoon depth to water table and the specific yield) is not a correct approach. What is important is to reckon with the fact that the storage space available in the aquifer for receiving the incoming flows keeps fluctuating from year to year. If a drought year is followed by an excessively wet year, it is quite likely that the rains produce a large amount of recharge, as the pre-monsoon depth to water level is likely to be high creating sufficient storage space. Second: while planning recharge schemes in the hard rock areas, it is important that the location chosen offers good infiltration (by virtue of the presence of dykes, lineaments, fractures, or alluvial deposits like in the valley portions), and the aquifer is dewatered before the onset of monsoon to create sufficient space for receiving the infiltrating water.

References

Alhassoun R (2011) Studies on factors affecting the infiltration capacity of agricultural soils. Dissertation report, Technischen Universität Carolo-Wilhelmina, Braunschweig

Arya S, Vennila G, Subramani T (2018) Spatial and seasonal variation of groundwater levels in Vattamalaikarai River basin, Tamil Nadu, India—a study using GIS and GPS. Indian J Geo Mar Sci 47(6):1749–1753

Bharathkumar L, Mohammed-Aslam MA (2018) Long-term trend analysis of water-level response to rainfall of Gulbarga watershed, Karnataka, India, in basaltic terrain: hydrogeological environmental appraisal in arid region. Appl Water Sci 8(4):1–9

Bhuiyan C, Singh RP, Flügel WA (2009) Modelling of ground water recharge-potential in the hard-rock Aravalli terrain, India: a GIS approach. Environ Earth Sci 59(4):929–938

CGWB (2021) National compilation on dynamic ground water resources of India, 2020. Central Ground Water Board (CGWB), Department of Water Resources, River Development & Ganga Rejuvenation Ministry of Jal Shakti, Government of India, Faridabad

Chandra S, Ahmed S, Ram A, Dewandel B (2008) Estimation of hard rock aquifers hydraulic conductivity from geoelectrical measurements: a theoretical development with field application. J Hydrol 357(3–4):218–227

Chatterjee R, Ray RK (2014) Assessment of groundwater resources: a review of international practices. Central Ground Water Board, Ministry of Water Resources, Govt. of India, Faridabad

Chen XY, Zhang K, Chao LJ, Liu ZY, Du YH, Xu Q (2021) Quantifying natural recharge characteristics of shallow aquifers in groundwater overexploitation zone of North China. Water Sci Eng. https://doi.org/10.1016/j.wse.2021.07.001

Deshpande S, Bassi N, Kumar MD, Kabir Y (2016) Reducing vulnerability to climate variability: forecasting droughts in Vidarbha region of Maharashtra, western India. In: Kumar MD, James AJ, Kabir Y (eds) Rural water systems for multiple uses and livelihood security. Elsevier Science, Amsterdam, pp 183–202

Dhar A, Sahoo S, Dey S, Sahoo M (2014) Evaluation of recharge and groundwater dynamics of a shallow alluvial aquifer in central Ganga basin, Kanpur (India). Nat Resour Res 23(4):409–422

Guhathakurta P, Saji E (2013) Detecting changes in rainfall pattern and seasonality index vis-à-vis increasing water scarcity in Maharashtra. J Earth Syst Sci 122(3):639–649

Healy RW, Cook PG (2002) Using groundwater levels to estimate recharge. Hydrogeol J 10(1):91–109

Kumar MD, Singh OP (2008) How serious are groundwater over-exploitation problems in India?: a fresh investigation into an old issue. In: Kumar MD (ed) Managing water in the face of growing scarcity, inequity and declining returns: exploring fresh approaches, vol 1. Proceedings of the 7th IWMI-TATA annual partners meet, ICRISAT, Hyderabad, 2–4 Apr 2008

Kumar TJ, Balasubramanian A, Kumar RS, Dushiyanthan C, Thiruneelakandan B, Suresh R, Karthikeyan K, Davidraju D (2016) Assessment of groundwater potential based on aquifer properties of hard rock terrain in the Chittar-Uppodai watershed, Tamil Nadu, India. Appl Water Sci 6(2):179–186

Machiwal D, Singh PK, Yadav KK (2017) Estimating aquifer properties and distributed groundwater recharge in a hard-rock catchment of Udaipur, India. Appl Water Sci 7(6):3157–3172

Maggirwar BC, Umrikar BN (2011) Influence of various factors on the fluctuation of groundwater level in hard rock terrain and its importance in the assessment of groundwater. J Geol Min Res 3(11):305–317

Maréchal JC, Dewandel B, Subrahmanyam K, Torri R (2008) Various pumping tests and methods for evaluation of hydraulic properties in fractured hard rock aquifers. In: Ahmed S, Jayakumar R, Salih A (eds) Groundwater dynamics in hard rock aquifers: sustainable management and optimal monitoring network design. Springer, Dordrecht, pp 100–111

Rajaveni SP, Brindha K, Elango L (2017) Geological and geomorphological controls on groundwater occurrence in a hard rock region. Appl Water Sci 7(3):1377–1389

Small D, Islam S, Vogel RM (2006) Trends in precipitation and streamflow in the eastern US: paradox or perception? Geophys Res Lett 33(3):1–4

Varalakshmi V, Rao, BV, SuriNaidu L, Tejaswini M (2014) Groundwater flow modeling of a hard rock aquifer: case study. J Hydro Eng, 2014 (19):877–886

Xiao J, Moody A (2004) Photosynthetic activity of US biomes: responses to the spatial variability and seasonality of precipitation and temperature. Glob Change Biol 10(4):437–451

Chapter 3
Factors Influencing the Performance of Rural Water Supply Schemes: Analysis from Maharashtra

3.1 Introduction

With the advent of handpumps introduced in the late 1960s and its successful introduction in remote rural areas of the country, India had made a strategic move to invest in groundwater-based schemes for rural water supply. There are three reasons for this: (i) the low capital cost of building the schemes to meet the localized water needs and a very low gestation period; (ii) the low level of technical sophistication of the system which meant that the local arms of the government (like the *Panchayats*) would operate and maintain them; and, (iii) the ease in finding source water to meet the low volumes of water demand. In the initial years, sustainability of the source water never appeared to be a concern, and it was thought that a sufficient amount of groundwater could be obtained locally to meet the low domestic demand.

With the passing of time, however, finding an adequate quantum of water in the wells to meet the rural drinking water requirements on annual basis proved to be a mounting task. With the intensification of groundwater withdrawal for irrigation in the semiarid and arid areas, the drinking water wells either started yielding less or in many cases began to dry up during the summer months. The efforts subsequently shifted to scientific prospecting of groundwater—in search of good aquifers and the right locations for drilling wells. This 'techno-centric' approach was also supported by a cadre of geologists who had a poor understanding of the physical and socio-economic aspects of agricultural water use.

The real magnitude of the resource sustainability challenges was not understood. While it was clear that irrigation claimed a significant share of the water in the aquifers in these regions, the chorus grew for stringent groundwater regulation. Such a reaction was more visible in states like Maharashtra, where rural water supply was dominated by wells. It was thought that a groundwater regulation (with strictly enforced norms on spacing and depth of irrigation wells) would help reduce the 'negative externalities' induced by irrigation pumping on drinking water wells. However, no Indian state could come up with a regulation to check groundwater over-development and protect the drinking water sources that was both technically meaningful and socially

M. Dinesh Kumar et al., *Drinking Water Security in Rural India*, Water Resources Development and Management, https://doi.org/10.1007/978-981-16-9198-0_3

acceptable. The 'groundwater acts' that were passed in the state legislatures were 'tooth-less' and devoid of substance. Here again, it should be noted that those who worked from behind the scene for drafting such regulations were 'pure geologists' who had little appreciation of the socio-economic dynamics in the villages. Basically, stopping or reducing irrigation water withdrawal meant that the farmers' livelihood is compromised.

At the same time, there are hundreds of thousands of drinking water wells and handpumps that survived the onslaught by irrigators in many regions and localities of the country that included areas that are traditionally known for widespread well failures such as hard rock areas with poor natural groundwater potential. Clearly, the factors that determine the sustainability of groundwater-based drinking water sources were not fully understood. Because of this, a lot of effort is also being made to invest in schemes to recharge groundwater locally using the monsoon run-off under various government schemes, with little appreciation of the dynamics of groundwater use. The essential point to recognize is that most of the agricultural water demand in the semiarid and arid areas is during the winter season, and this demand exceeds the total available resource, whereas the domestic water demand is throughout the year. This basically means that any additional recharge to the local aquifers only benefits the irrigators, who would in turn increase their use of water by expanding the cropped area.

In this chapter, we present the results of the investigation into the factors that explain the varying performance of rural water supply schemes using data from Maharashtra. Building on the analysis presented in the previous chapter that looked at the impact of various factors on groundwater level fluctuations during monsoon, the analysis presented in this chapter looks at the potential influence of additional factors that are physical (geomorphology, climate), technical (dependence on groundwater, extent of gravity surface irrigation) and socio-economic (cropped area and irrigated area) in nature.

The subsequent section of this chapter provides a review of available scientific literature on the theme of sustainability of groundwater-based schemes, with special reference to India. This is followed by the description of a statistical model developed and validated by us to explain the varying performance of water supply schemes in Maharashtra under different physical environments. The state's widely varying environmental (rainfall, climate, geohydrology) and socio-economic conditions made it the ideal place for testing the validity of the model. An assessment of the likely success of rural water supply schemes in each district was done on the basis of how the districts fared on these parameters. Based on this comparative assessment, four typologies of districts are identified. Each typology required a certain type of intervention to improve the sustainability of water supply schemes. The chapter also explains the practical and policy relevance of the model's results in rural water supply planning.

3.2 Review of Literature

In India, currently, the majority of the water supply schemes are based on groundwater. The rural households use water supplied from the schemes for many productive purposes (such as for homesteads and small business activity) in addition to the intended domestic uses (Smits et al. 2010; Kumar et al. 2016). Therefore, at the aggregate level, the water supply levels maintained by the schemes, which is as per traditional norms of per capita domestic needs, is not able to fully meet the demands in most instances. Further, the same groundwater resource is in demand for many competing uses that include irrigation and industrial purposes. Therefore, the sustainability of drinking water supplies based on such sources would depend on many factors relating to the availability, demand, and use of water.

As demonstrated in Chap. 2, on the availability side, the most crucial factors for the hard rock areas are the ground conditions that determine the potential infiltration from rainfall, and the storage space available in the aquifer to accommodate the infiltrating water. The combined effect of these variables will be on the amount of water that remains in the aquifer to sustain the drinking water sources which ultimately determine the performance of the scheme (wells, handpumps, etc.) that tap groundwater. However, a portion of the water that is stored in the aquifer during the monsoon (and that is reflected in the water level fluctuations during the season) is discharged into the streams as the water level in the streams starts receding after the monsoon. This discharge is largely governed by the geomorphological conditions that determine the groundwater flow gradient, geohydrology (especially transmissivity of the aquifer) and the interface of the aquifer with the streams. In hilly and mountainous areas, the fraction of the monsoon recharge that is discharged into the stream (also known as lean season base flow) is generally very high (Deshpande et al. 2016; Winter et al. 1998).

An important factor that has received the least attention in the debate on drinking water security is the role of surface irrigation on the sustainability of drinking water wells. As discussed earlier, the past efforts to improve the sustainability of water supply from wells in the hard rock areas (like in the Saurashtra region of Gujarat, Marathwada, and Vidarbha regions of Maharashtra) had focussed on building artificial recharge schemes to put monsoon run-off into the aquifers. However, the strategy has not been effective due to the fact that during a good monsoon (when run-off is normally available from the local catchments for capturing), the hard rock aquifers get fully replenished with very little extra space in the aquifer to receive additional infiltration caused by the recharge structures (Kumar et al. 2006). For recharge to be effective and to have a positive impact on the yield of drinking water wells, infiltration should occur during the lean season because during this season there is no natural recharge from rainfall and the aquifers are already empty due to excessive pumping for irrigation. Since irrigation schemes generally supply water through canals during winter and sometimes during summer, effective infiltration happens through canal seepage (Watt 2008) and return flows from irrigated fields to affect replenishment of groundwater and rejuvenation of drinking water wells (Kumar et al. 2014).

Research studies undertaken in the Murray basin of Australia and Indus basin in Pakistan, both having arid climate and deep water table conditions, showed a substantial rise in water table conditions after the introduction of canal water for irrigation. Such changes happen due to reduced pumping of groundwater with the introduction of canal water and gradual increase in moisture storage in the unsaturated zone, which increase the unsaturated hydraulic conductivity of the soil media and the increase in moisture pressure (hydraulic) gradient of the soil (Watt 2008). While the first factor reduces or keeps the vertical distance for movement of recharge water, the second and third factors increase the rate of vertical movement of soil water (source: based on Richards Equation and van Ganuchten Equation as cited in Watt 2008, pp. 101–102). Kumar et al. (2014), which covered the Sardar Sarovar Project command, also reported dilution of chemical contaminants such as salts in groundwater as a result of return flows from gravity irrigation.

As regards the use, the most important is the demand for water in irrigation and the availability of water from other sources (especially surface water) to meet those demands. Climate is an important factor that influences the demand for irrigation water (Kumar et al. 2021), along with the proportion of the geographical area under cultivation. The higher the aridity, the higher would be the demand for irrigation water per unit area. Higher the effective rainfall, lower would be the irrigation water demand per unit area. Additionally, an increase in area under cultivation (as a proportion of the total geographical area) would also increase the demand for irrigation water, with climatic conditions remaining the same.

Thus, any adverse situation resulting from either reduced availability of water or increased withdrawal of water has the potential to influence the functioning of the scheme. It was found that, overall, in about 10–15% of the states in India, habitations covered by the groundwater-based scheme suffer ‘slip-back’ annually (Chaudhuri et al. 2020). ‘Slip-back’ is a situation when a habitation covered by a drinking water scheme fails to receive the water supply as per the existing norm. For groundwater-based schemes, the water supply norm is set at 40 l per capita per day.

The proportion of habitations getting ‘slip-back’ to non-functional water supply scheme or no-source category is more frequent in the hard rock areas. In Andhra Pradesh, over the years more than 30% of the habitations dependent mostly on groundwater-based schemes ‘slip-back’ to partially covered category (Reddy et al. 2010), meaning the supply of water was less than the accepted norm of 40 lpcd. Similarly, almost 42% of the 48 surveyed schemes in 2014–15 were either non-functioning or only partially functioning in Maharashtra (Sakthivel et al. 2015).

The lack of adequate quantity of water supply and drying up of the source are identified as among the major reasons for slippage of the rural habitations (Reddy et al. 2010). Analysis of data from 677 ‘slip-back’ habitations in the Konkan division of Maharashtra revealed that about 32% of the habitation had ‘slip-back’ due to drying up of the source. In the remaining 68% of the habitations, schemes were not able to perform satisfactorily (due to various reasons) even when the design life was yet to complete. Less supply at the delivery point and improper functioning of the old scheme were found to be the other two main reasons for the failure of the schemes in the region (Table 3.1). Similarly, in Amravati, Aurangabad, Nagpur, and

Table 3.1 Reasons for failure of rural water supply scheme in Konan division, Maharashtra

Sr. No.	Slip back reason for habitation	Failure type	No. of habitations	% of total
1	Quality problem reported	Planning/O&M	5	0.7
2	Shortage of electricity	Planning	13	1.9
3	Poor O&M	O&M	22	3.25
4	Old scheme is outdated	O&M	201	29.5
5	Drying of sources	Planning	217	32.0
6	Less supply at delivery point	O&M	219	32.3

Source Based on data presented in IRAP et al. (2018)

Pune, 50% of the non-functional water supply schemes were due to drying up of the groundwater-based source (Sakthivel et al. 2015).

Seasonal shortage of water is another reason for the temporary failure of groundwater-based supply schemes in the hard rock areas (Reddy et al. 2010). Further, it has been observed that inadequate monetary allocation for interventions related to source sustainability is one of the main causes of widespread slippage in the rural water supply services. For instance, in Andhra Pradesh, the government expenditure was almost exclusively for developing water supply infrastructure while other important components like planning and designing of the scheme, maintenance of infrastructure, interventions to protect water supply source (source sustainability), and interventions to improve water quality receive little or no allocation (Reddy et al. 2012). As a result, the water supply schemes are often built at locations without appropriate assessment studies of future sustainability; even the yield tests were rarely performed for the prospective sources (Chaudhuri et al. 2020).

The foregoing discussion indicates that the sustainability of the water supply scheme is influenced mainly by physical (choice of water supply source and technology used for supply), environmental (climate, geology, geo-morphology), socio-economic (water demand for various domestic and productive uses), and financial factors.

3.3 Methodology

To determine the sustainability of the scheme, a multiple regression analysis was performed by considering the number of tankers supplied during summer per '000 population as a dependent variable and the water demand for irrigation per unit geographical area, effective utilizable recharge per unit area, the proportion of the total geographical area under surface irrigation, the extent of habitation covered by groundwater-based schemes for water supply and storage space in the aquifer as independent variables. The variables used in the regression analysis were adjusted for calibration of the regression-based model to get the strong coefficient.

To compute all the variables used for the regression analysis, district-wise data were collected for the years 2016, 2018, and 2019. These include rainfall, total geographical area, net sown area, surface irrigated area, rural population, number of habitations covered by groundwater-based schemes, number of tankers supplied, and pre- and post-monsoon groundwater levels. Maps on geo-morphology and the one showing priority areas for artificial recharge (prepared by Groundwater Surveys and Development Agency, Government of Maharashtra) were referred for choosing the recharge coefficient and the specific yield of the aquifer. The data for 2017 were incomplete, and thus not considered for the analysis. The approach used to compute various variables is discussed in the relevant section of the chapter.

The output model (result) from the regression analysis was used to identify factors that determine the success of the groundwater-based schemes. It analysed the impact of each of the independent variables on the dependence of the tanker water supply and shortlisted those that reduce the dependence on the tanker water supply during the lean season. Various districts in Maharashtra were ranked on the identified factors, and the bottom 10 districts under each attribute were categorized into distinct typologies, each requiring a certain type of intervention to improve the sustainability of water supply schemes.

3.4 Scheme Failure and Tanker Water Supply in Rural Areas

To determine the success or failure of a rural water supply scheme, data on the yield of drinking water wells and handpumps (outcome indicator) and data on the supply and demand related variables (potential determinants of performance of schemes) are needed. However, such data sets are not readily available and it would require an enormous amount of effort to generate them. Therefore, a more pragmatic approach was followed by considering a simple indicator of sustainability of rural water supply, i.e. the extent of dependence on tanker water supply during lean season. It is a common phenomenon that whenever groundwater-based schemes go dry (during summer months and during droughts), the village-level community-based institutions (such as village Panchayats) obtain tanker water supply provided by the government. Hence, it was examined whether the high degree of dependence on groundwater-based sources has any negative bearing on the sustainability of rural water supply schemes. For this, the data on the extent of tanker water supply in different districts of Maharashtra during 2016, 2018 and 2019 were analysed. The data for 2017 were incomplete and thus not considered for the analysis. Since the population varies from district to district, the total number of tankers supplying water was divided by the population in thousand to obtain tankers per '000 population in order to normalize the figures.

The data on tankers per '000 people are presented in Fig. 3.1. Overall, the extent of tanker water dependence was found to be highest in 2019. Across districts, the

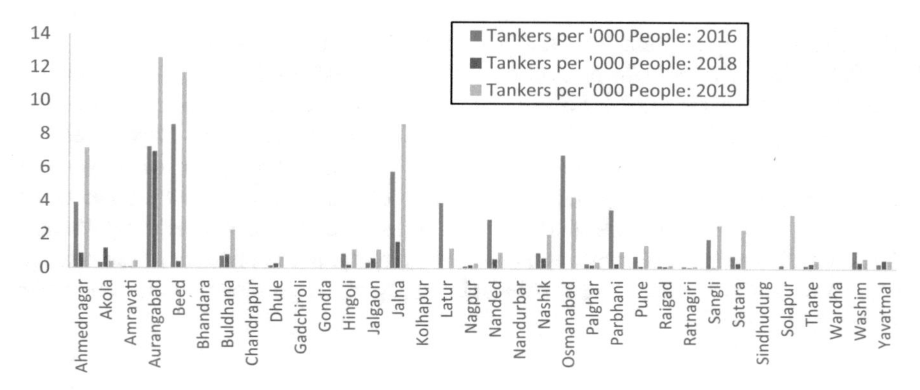

Fig. 3.1 Tanker water supply in districts of Maharashtra in 2016, 2018, and 2019. Authors' analysis based on data from the Water Resources Department, Government of Maharashtra

highest incidence of tanker water dependence was found in Aurangabad district, followed by Beed, Jalna, Ahmednagar, and Osmanabad.

Further, a frequency analysis was done with the proportion of rural habitations covered by groundwater-based schemes and the extent of use of tanker water supply. A total of 34 districts from the state were chosen for the analysis. Mumbai city and Mumbai suburban districts were not considered for the analysis as they are urban. The districts were put into three categories: those where the proportion of rural habitations depending on groundwater is less than 80%, those where the proportion of rural habitations depending on groundwater is in the range of 80–90%, and the last one in which more than 90% of rural habitations are served by groundwater. The results are presented in Table 3.2.

It was found that in those districts where the proportion of habitations covered by groundwater-based schemes is less than 80% (there are 5 such districts and the actual % ranges from 34 to 79), the tanker water supply (per 1000 persons) is very low—0.27, 0.28, and 1.081, respectively, in 2016, 2018, and 2019. Against this, for districts where the proportion of habitations covered by groundwater-based schemes is in the range of 80 and 90% (there are 11 such districts), the tanker water supply per

Table 3.2 Water supply scheme characteristics and dependence on tanker water supply

Proportion of habitations depending on GW-based schemes for water supply	Total no. of districts in the category	No. of tankers supplying water per 1000 persons in		
		2016	*2018*	*2019*
0.0–80.0%	05	0.27	0.28	1.08
80.01–90.00	11	1.39	0.31	1.55
90.01–99.45%	18	1.95	0.67	2.51

Source Authors' analysis based on data from Ministry of Jal Shakti, GoI

1000 persons is quite high—1.39, 0.309, and 1.55, respectively, in 2016, 2018, and 2019. For districts where the proportion of habitations covered by groundwater-based schemes is in the range of 90% and above (there are 18 such districts), the tanker water supply per 1000 persons is quite high—1.95, 0.67, and 2.505, respectively, in 2016, 2018, and 2019. The above analysis suggests that the higher the proportion of habitations covered by groundwater-based schemes the higher the chances of water supply becoming undependable. The differences are statistically significant. The analysis brings out one phenomenon very clearly. In the districts which heavily depend on surface sources for rural water supply, the dependence on tanker water supply is quite low.

3.5 Influence of Irrigation Water Demand on Sustainability of Schemes: Theoretical Perspective

In a normal rainfall year in Indian conditions, demand for irrigation water mainly starts during winters when most of the soil moisture is used by the crops taken during the monsoon season. Many hard rock areas have semiarid to arid climate which is characterized by low rainfall and high evaporation and transpiration rates. Further, as only limited soil moisture is available during non-monsoon months, crops taken during winters in such areas have high irrigation water demand.

In some parts of such areas, the terrain is undulating making the supply of water through gravity-based surface irrigation systems a difficult proposition. As a result, farmers look for groundwater as a dependable source of irrigation. Over the years, both the number of irrigation wells and their depth have increased.

However, as the same aquifer is shared by many farmers, the overall abstraction often exceeds the annual utilizable groundwater recharge (Anantha 2009) and the majority of utilizable groundwater is withdrawn for irrigating winter crops. Thus, the wells dry up before the onset of summer leading to seasonal scarcity of groundwater. It has been observed that due to high dependence on groundwater for irrigation, water yield of the wells has decreased (Kumar and Singh 2008) and the rate of wells failure has increased (Kumar 2007; Kumar and Singh 2008; Bassi et al. 2008) in the hard rock areas. Even the deeper bore wells have a poor yield in such formations. As high as 60% of about 40,000 deep bore wells (that are in use) in Maharashtra were not able to utilize their potential due to poor discharge (Kumar et al. 2010).

Under such a scenario, the functioning of groundwater-based rural water supply schemes gets affected adversely and they fail to provide water during summers and become unsustainable. Every year, newspapers are flooded with the reports on drinking water crisis during the summer months in the hard rock areas of India.

3.6 Impact of Gravity Irrigation on Rural Water Supply Schemes: Theory and Empirical Evidence

Regions with a significant proportion of cropped area under gravity-based canal irrigation have less pressure on groundwater to meet the irrigation demand. Most of the canals in India run during winters when there is a need for irrigating crops. Though there is a conjunctive use of water in the canal command areas, wells in such settings get recharge benefit during the non-monsoon months from the canal seepage losses.

In the semiarid alluvial plains of north-west India, the contribution of canal seepage to total groundwater recharge was estimated to vary from 5 to 83% depending on the depth of the aquifer (Joshi et al. 2020). The benefit to groundwater recharge in the command areas is also influenced depending on whether the canals are lined or unlined. In the hard rock catchment area of the Cauvery River basin in southern India, the contribution of the lined canals to groundwater recharge was up to 20%, whereas it was up to 40% in the case of the unlined canals (Mirudhula 2014).

As of 2018–19, the majority of gravity canal-based surface irrigated area in Maharashtra is confined to the districts of Ahmednagar, Pune, Solapur, Satara, Sangli, and Kolhapur. These districts together constitute about 53% of the total 2.06 million ha of surface irrigated area in the state. Further, these districts along with Gondia and Bhandara have a higher proportion of arable land under surface irrigation in comparison with other districts (Fig. 3.2). Except for Ahmednagar, these are also the districts where the incidence of tanker dependence is among the lowest in 2016 and 2018 (refer to Fig. 3.1). In fact, in Kolhapur, Gondia, and Bhandara, no tanker water supply was requested in 2016 and 2018.

In Aurangabad, Beed, Jalna, and Osmanabad, which have the highest incidence of tanker water dependence (refer to Fig. 3.1), the proportion of total arable land under the gravity-based surface irrigation is among the lowest and in most cases below

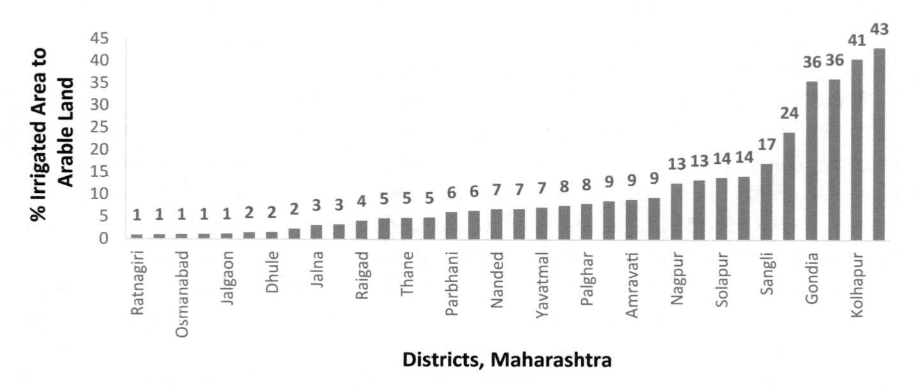

Fig. 3.2 Extent of arable land under surface irrigation. *Source* Authors' analysis using data from Water Resource Department, Government of Maharashtra

4% (Fig. 3.2). Thus, it can be interpreted that in hard rock areas in the command of gravity-based surface irrigation systems, the groundwater remains available for a longer duration (in some instances, lasting the whole year) than in the non-command areas.

3.7 The Model Explaining the Varying Performance of Rural Water Supply Schemes

It was observed that for many districts with coverage of groundwater schemes touching a very high extent—going beyond 90%—the use of tanker water supply is quite low. Examples are Gadchiroli, Nandurbar, Nagpur, Washim, and Ratnagiri. Therefore, merely having a situation of a high degree of dependence on groundwater does not mean that the water supply schemes will have low dependability and that the villages will have to depend more on tanker water supply. There are many factors that determine the performance of groundwater-based drinking water supply schemes. The condition in various districts of the state with respect to those factors needs to be understood.

Firstly, the rainfall and geomorphological conditions keep varying from district to district (even within divisions); significantly, both might be probably influencing the yield of drinking water wells during the summer season. Flat topography and permeable soils will provide favourable conditions for the rainwater to infiltrate. On the other hand, a higher quantum of rainfall not only increases the recharge to aquifers but also reduces the demand for irrigation which is the largest user of water in almost all districts, except the highly urbanized districts of Mumbai and New Mumbai.

Secondly, the aquifer conditions also vary in the state though 90% of the state's geographical area is underlain by basalt. In the Konkan region, with very high rainfall and more favourable geology—with laterite and alluvial sands along with basalt formations—groundwater storage potential is good. In the Tapi basin area (extending over part of five districts, viz. Akola, Amravati, Dhule, Jalgaon, and Buldhana) also, due to alluvial deposits, the groundwater potential is good.

Thirdly, as we have mentioned earlier, surface water availability for irrigation can be a very important determinant influencing the sustainability of groundwater-based drinking water schemes. Since such data are not available readily at the district level, a proxy variable called the proportion of geographical area under surface irrigation was used. Higher the proportion of the area under surface irrigation, lower will be the dependence on groundwater for irrigation and higher would be the recharge to groundwater, particularly during the non-monsoon period. Similarly, water demand for agriculture that is directly met from groundwater can be another important variable. A robust coefficient that can truly represent the intensity of demand for groundwater for irrigation at the aggregate level is the multiple of the net sown area as a proportion of the total geographical area (NSA/GA) and the difference between the reference evapotranspiration (ETo) and effective rainfall (P_e). It can be represented

Table 3.3 Results of multi-variate analysis

	Coefficients	Standard error	*t* stat	*P* value
Intercept	2.67	3.670	0.73	0.472
Storage available in the aquifer	−3.046	1.050	−2.90	0.007
Irrigation demand from year (mm)	0.00689	0.00165	4.17	0.000
Proportion of habitations covered by groundwater	−0.0187	0.0312	−0.60	0.554
Proportion of geographical area under surface irrigation	−14.45	7.824	−1.85	0.076
Eff. utilizable recharge/unit area	−0.02245	0.0256	−0.87	0.39

Source Authors' own analysis

as $NSA \times (ETo - P_e)/GA$. Here, the ETo and P_e values have to be for the entire length of the cropping season. If there are standing crops for 240 days in a year in a district, the ETo and P_e values shall be estimated for that many days.

To analyse the effect of rainfall, climate, proportion of area under cultivation and surface water availability for irrigation, a multi-variate analysis was performed with the following variables: (1) the water demand for irrigation per unit geographical area expressed as $NSA \times (ETo - P_e)/GA$; (2) effective utilizable recharge per unit area, which is a multiple of the rainfall, recharge coefficient chosen for the district on the basis of geomorphology (used in the analysis in Chap. 2), and a coefficient called 'utilizable recharge fraction' to arrive at the recharge actually available for use during the year, or to factor out the baseflow from the aquifer into streams during the non-monsoon period[1]; (3) the proportion of the total geographical area under surface irrigation (in %); (4) extent of habitation covered by groundwater-based schemes for water supply (in %); and (5) multiple of specific yield of the aquifer and pre-monsoon depth to water levels (storage space in the aquifer, i.e. $STORAGE_{AQUIFER}$), a factor that is found to have a significant influence on the total recharge from monsoon as per the analysis presented in Chap. 2. The values of various coefficients used in the analysis for each district are presented in Table 3.2.

The regression analysis shows that all these factors influence the extent of dependence on tanker water supply, which is an indicator of the failure of groundwater-based schemes. The R-square value was estimated to be 0.51 for 2016. The regression equation is provided below, and the results are presented in Table 3.3.

$$\begin{aligned} WS_{TANKER} = {} & 2.67 - 14.45 \times IRR_{SW} - 0.022 \times RECH_{GW} \\ & - 3.046 \times STORAGE_{AQUIFER} - 0.0187 \times HHAB_{GWS} \\ & + 0.00685 \times NSA \times (ETo - P_e)/GA \end{aligned}$$

[1] For hilly areas with very high rainfall such as Ratnagiri, Raigarh and Sindhudurg, the value for utilizable recharge fraction considered is 0.20. For hilly and undulating areas (Chandrapur, Pune, Ahmednagar, etc.) the value considered is 0.50. For remaining areas, which are part of the Deccan plateau, the fraction considered is 0.80. A value of 0.20 means that only 20% of the recharge available during the monsoon will be utilizable.

where WS_{TANKER} is the tanker water supply, IRR_{SW} is the proportion of the total geographical area under surface irrigation, $RECH_{GW}$ is effective utilizable recharge per unit area, $STORAGE_{AQUIFER}$ is the aquifer storage space, and $HHAB_{GWS}$ is the proportion of habitations covered by the groundwater-based water supply schemes.

As per the results of the multi-variate regression (refer to Table 3.3), the coefficients are negative for four variables, viz. storage space available in the aquifer; percentage of geographical area covered by surface irrigation; proportion of habitations covered by groundwater-based schemes; and effective utilizable groundwater recharge per unit geographical area. This trend is along the expected lines. It should also be noted that the level of significance for recharge per unit area and the proportion of habitations covered by groundwater are low. The level of significance is very high for storage space available in the aquifer, which is the multiple of the pre-monsoon depth to water level and specific yield of the aquifer, and high for 'percentage geographical area under surface irrigation'.

The coefficient is positive for one variable, i.e. intensity of groundwater demand for irrigation (with a very high level of significance), which again is along the expected lines. Under normal circumstances, in a situation like in Maharashtra, with hard rock aquifers in most parts, increasing demand for irrigation water (with a greater proportion of the geographical area under cultivation, more aridity, and intensive cultivation) would ideally lead to faster depletion of the aquifers, with increased dependence on tanker water supply for drinking water supply needs. Conversely, increasing the provision of surface water for meeting the irrigation needs will have positive effects, as shown by the model. First: it reduces the groundwater demand for irrigation. Second: it improves groundwater recharge through irrigation return flows. The regression equation suggests that with every 100 mm increase in irrigation demand the dependence on tanker water supply will increase by 0.68 tankers per 1000 people. Similarly, for every one m^3 increase in storage space in the aquifer per m^2 of the geographical area, the dependence on tanker water supply is likely to reduce by nearly 3 tankers per 1000 people.

3.8 Factors Determining the Success of Groundwater-Based Schemes

As we have explained, the easiest way to understand the conditions under which groundwater-based schemes succeed is to understand the conditions that reduce the dependence on tanker water supply during the lean season. In this regard, the analyses presented in the earlier sections suggest the following. In districts where the aquifer has enough storage space before the onset of monsoon (indicated by the multiple of pre-monsoon depth to water table and specific yield) and if the demand for water for irrigation is low (by virtue of low proportion of the area under cultivation and high rainfall and low aridity), then groundwater-based schemes will be successful. The natural recharge potential of groundwater (by virtue of favourable geomorphological conditions and good rainfall) also has some positive influence on the sustainability of groundwater-based schemes.

Here again, we need to reckon with the fact that the groundwater recharge potential of a district can decline considerably and the pressure on groundwater to meet irrigation water demand can also go up significantly when the rainfall is low in a particular year. The pattern of occurrence of the rainfall also will have some effect. The same amount of rainfall distributed over greater number of rainy days will produce a greater quantum of recharge than what would be produced if the same rainfall occurs on fewer rainy days. However, we have not considered its effect on recharge in the present model.

Multi-variate analysis also showed that the dependence on tanker water supply is likely to be very less in districts that get a considerable amount of surface water for irrigation, as indicated by a high proportion of the geographical area under canal irrigation. When surface irrigation becomes extensive, not only does the pressure on groundwater for irrigation reduces, but the recharge to groundwater during the lean season through irrigation return flows increases. So, even in districts underlain by hard rock aquifers, groundwater-based drinking water schemes are likely to sustain during summer months if there is considerable surface irrigation during the non-monsoon season.

Overall, it appears that for groundwater-based drinking water supply schemes in rural Maharashtra to be sustainable several conditions have to be met. They are favourable geo-hydrological environment (sufficient storage space in the aquifer before the onset of monsoon, due to high depth to water table and good specific yield of the aquifer); large extent of surface water-import for irrigation or very low irrigation water demand; and favourable geomorphology and rainfall that increase the utilizable recharge rates. The districts which score lowest on each of the four key attributes—sufficient recharge rate per unit area; adequate aquifer storage space; limited irrigation demand; and increasing the extent of surface irrigation—are given in Table 3.4.

Table 3.4 10 districts of Maharashtra with lowest scope on the four attributes

Recharge per unit area (mm)	Aquifer storage space (m)	Limited irrigation water demand (mm/year)	Proportion of area under surface water irrigation
Sangli	Chandrapur	Gadchiroli	Gadchiroli
Solapur	Gadchiroli	Gondiya	Gondiya
Satara	Kolhapur	Pune	Pune
Ahmednagar	Gondiya	Chandrapur	Chandrapur
Kolhapur	Nagpur	Nandurbar	Nandurbar
Nagpur	Bhandara	Satara	Satara
Aurangabad	Pune	Kolhapur	Kolhapur
Bhandara	Aurangabad	Bhandara	Bhandara
Beed	Satara	Nashik	Nashik
Pune	Yavatmal	Wardha	Wardha

Source Authors' own analysis based on secondary data

The information provided in Table 3.4 needs to be read and interpreted very carefully. Table 3.4 shows that Gadchiroli has the lowest score in terms of limited irrigation water demand. This basically means that the district has the highest irrigation demand (mm/year). Similarly, Sangli has the lowest score for 'Recharge per Unit area'. This means that the district has the lowest recharge rate (mm).

There are very few pockets in Maharashtra (only parts of certain districts) where the conditions are favourable for groundwater-based schemes for sustainable rural water supply. They include the tracts of the Konkan division in three districts, viz. Ratnagiri, Raigad, and Sindhudurg (which also receive very high rainfall), having laterite formations and alluvial deposits and certain pockets in the Amravati division, along the Tapi river valley extending from Dhule to Akola to Amravati to Jalgaon and Buldhana, which receive considerable recharge from the river itself. The rest of the areas either suffer from poor rainfall or poor recharge potential of aquifers or both. In western Maharashtra, though the rainfall is excessively high, the recharge potential is very poor due to the steep terrain and hard rock terrain which result in poor infiltration of the rainwater. Further, due to the steep groundwater flow gradient, the recharge during monsoon does not remain in the aquifer due to heavy groundwater discharge into streams that drain the region.

3.9 Conclusions and Policy Implications

The main conclusion emerging from the study is that there are very few areas in Maharashtra in terms of geographical extent where condition is favourable for provision of water supply through wells because of the unique physical (hydrological, geological, geohydrological, geomorphological, and climatic) and socio-economic conditions (demand for irrigation water and access to surface water to meet these demands) that determine the availability of water in the aquifers and demand for water across regions and seasons.

Comparing the situation in Maharashtra with that of India, it is apparent that there are large regions where the situation is unfavourable for managing the required quantity of water for rural water supply throughout the year, due to unfavourable conditions vis-à-vis groundwater availability and demand. Around 2/3rd of the country's geographical area is underlain by hard rock formations of various origins. Most of these regions also experience excessive demand for water for irrigation due to the presence of a large amount of arable land, low to medium rainfall and high aridity. Groundwater-based rural water supply schemes are increasingly becoming unsustainable in these regions (Kumar et al. 2021). Future investments for rural water supply in these regions will have to be based on more dependable surface water sources.

On the other hand, there are large regions in India where the condition is quite favourable vis-à-vis groundwater availability and overall pressure on the resource from various competing uses, with good availability and demands lesser than the availability. These are regions with rich alluvial aquifers, large stocks of groundwater over and above the annual recharge from precipitation, and relatively low pressure on groundwater for irrigation due to relatively high precipitation, lower aridity, and relatively good access to surface irrigation and therefore capable of ensuring year-round drinking water provisioning through wells. The entire Indo-Gangetic Plain is part of this area. In addition, there are alluvial tracts in other parts of the country too. Most of these regions rely on wells (shallow and deep tube wells) and handpumps for rural water supply. Failure of schemes due to inadequate supply and source drying is very rare in these regions. However, the schemes in some pockets of these areas would be increasingly facing challenges in the future due to groundwater contamination and pollution of various kinds and from various sources (Kumar et al. 2021). We will closely examine those challenges in Chap. 7.

References

Anantha KH (2009) Downward dividends of groundwater irrigation in hard rock areas of southern peninsular India. Working paper 225. The Institute for Social and Economic Change, Bangalore

Bassi N, Vijayshankar PS, Kumar MD (2008) Wells and ill-fare: impacts of well failures on cultivators in hard rock areas of Madhya Pradesh. In: Kumar MD (ed) Managing water in the face of growing scarcity, inequity and declining returns: exploring fresh approaches, vol 1. Proceedings of the 7th IWMI-TATA annual partners meet, ICRISAT, Hyderabad, 2–4 Apr 2008

Chaudhuri S, Roy M, McDonald LM, Emendack Y (2020) Water for all (*har ghar jal*): rural water supply services (RWSS) in India (2013–2018), challenges and opportunities. Int J Rural Manag 16(2):254–284

Deshpande S, Bassi N, Kumar MD, Kabir Y (2016) Reducing vulnerability to climate variability: forecasting droughts in Vidarbha region of Maharashtra, western India. In: Kumar MD, James AJ, Kabir Y (eds) Rural water systems for multiple uses and livelihood security. Elsevier Science, Amsterdam, pp 183–202

IRAP, CTARA, UNICEF (2018) Compendium of training materials for the capacity building of the faculty and students of engineering colleges on improving the performance of rural water supply and sanitation sector in Maharashtra: under the Unnat Maharashtra Abhiyan (UMA). UNICEF, Mumbai

Joshi SK, Rai SP, Sinha R (2020) Understanding groundwater recharge processes in Sutlej-Yamuna plain in northwest India using isotopic approach. Geol Soc Lond 507. https://doi.org/10.1144/SP507-2020-174

Kumar MD (2007) Groundwater management in India: physical, institutional and policy alternatives. Sage, India

Kumar MD, Singh OP (2008) How serious are groundwater over-exploitation problems in India? A fresh investigation into an old issue. In: Kumar MD (ed) Managing water in the face of growing scarcity, inequity and declining returns: exploring fresh approaches, vol 1. Proceedings of the 7th IWMI-TATA annual partners meet, ICRISAT, Hyderabad, 2–4 Apr 2008

Kumar MD, Ghosh S, Patel A, Singh OP, Ravindranath R (2006) Rainwater harvesting in India: some critical issues for basin planning and research. Land Use Water Resour Res 6:1–17

Kumar MD, Narayanamoorthy A, Sivamohan MVK (2010) Pampered views and parrot talks: in the cause of well irrigation in India. Occasional paper 1. Institute for Resource Analysis and Policy, Hyderabad

Kumar MD, Jagadeesan S, Sivamohan MVK (2014) Positive externalities of irrigation from the Sardar Sarovar Project for farm production and domestic water supply. Int J Water Resour Dev 30(1):91–109

Kumar MD, Kabir Y, James AJ (eds) (2016) Rural water systems for multiple uses and livelihood security. Elsevier, Netherlands, UK and USA
Kumar MD, Bassi N, Hemani R, Kabir Y (2021) Managing climate-induced water stress across agroecologies of India: options and strategies. In: Kumar MD, Kabir Y, Hemani R, Bassi N (eds) Management of irrigation and water supply under climatic extremes: empirical analysis and policy lessons from India. Springer Nature, Switzerland, pp 313–354
Mirudhula K (2014) Impact of lined/unlined canal on groundwater recharge in the lower Bhavani Basin. Int J Innov Appl Stud 9(4):1818
Reddy VR, Rammohan Rao MS, Venkataswamy M (2010) 'Slippage': the bane of drinking water and sanitation sector (a study of extent and causes in rural Andhra Pradesh). WASHCost India-CESS working papers. Hyderabad
Reddy VR, Jayakumar N, Venkataswamy M, Snehalatha M, Batchelor C (2012) Life-cycle costs approach (LCCA) for sustainable water service delivery: a study in rural Andhra Pradesh, India. J Water Sanit Hyg Dev 2(4):279–290
Sakthivel SR, Dhar NS, Godkhe A, Gore G (2015) Status of rural water supply in Maharashtra. Tata Institute of Social Sciences, Mumbai
Smits S, Van Koppen B, Moriarty P, Butterworth J (2010) Multiple-use services as alternative to rural water supply services: a characterisation of the approach. Water Altern 3(1):102–121
Watt J (2008) The effect of irrigation on surface-ground water interactions: quantifying time dependent spatial dynamics in irrigation systems. Thesis submitted to Charles Sturt University for the degree of Doctor of Philosophy, School of Environmental Sciences, Faculty of Sciences, Charles Sturt University
Winter TC, Harvey JW, Franke OL, Alley WM (1998) Ground water and surface water a single resource. US geological survey circular 1139. Denver, CO

Chapter 4
Studying the Performance of Rural Water Supply Schemes in Different Geological Settings

4.1 Background

Till the mid-1990s, the whole approach to the provision of rural water supply in India was 'supply-driven'. The approach had an emphasis on meeting simplistic norms of per capita water requirement and construction and creation of water supply infrastructure and the related assets. However, there was little progress in terms of the number of villages and habitations covered under such an approach. To address this concern, a policy shift was introduced in 1999 towards a 'demand-driven' approach where community participation and decentralization of powers for implementing and operating drinking water supply were made the essential points. The government role was restricted to that of a facilitator. The basic idea was to take water supply to rural areas that remained unserved for long.

Since the rural community was made the major stakeholder for planning and implementing the schemes under the demand-driven approach, groundwater-based schemes supplying water to single habitation or a village became the preferred choice as the community found it easy to build, operate, and maintain such schemes. This demand-driven participatory approach was lauded, and suggestions were made to continue the same for better outcomes in terms of community satisfaction with the scheme design and improvement in access to water (Isham and Kähkönen 2002; Prokopy 2005). Though the choice of opting for more convenient groundwater-based schemes did help significantly in terms of improving the water supply coverage (Moriarty et al. 2013), such schemes were unable to provide year-round drinking water security with an increasing number of habitations slip-back to the 'no-source category'. One of the main reasons was the poor sustainability of water supplies owing to an unsustainable resource base (Reddy et al. 2010; Bassi and Kabir 2016).

Further, little information is available on how effective the expenditure has been in providing safe water to the rural population (World Bank 2008). There is hardly any analysis of the cost of water supply schemes, cost recovery and subsidies, and the impact of technology choice and institutional arrangements on the level of service using the field data. Pattanayak et al. (2007) observed that there were only a few

M. Dinesh Kumar et al., *Drinking Water Security in Rural India*, Water Resources Development and Management, https://doi.org/10.1007/978-981-16-9198-0_4

rigorous scientific impact evaluations showing the effectiveness of rural water supply policies in delivering many of the desired outcomes. Their analysis showed that such policies are complex with multiple objectives; use inputs from multiple sectors; provide a variety of services (water supply, water quality, sanitation, sewerage, and hygiene) using a variety of types of delivery (public intervention, private interventions, public–private partnerships, decentralized delivery, expansion, or rehabilitation); and generate effects in multiple sectors (water, environment, health, labour). The fact that the communities are mostly dependent on multiple sources of water supply (Kumar et al. 2016), including informal sources, often makes the evaluation of policy impacts complicated.

As an outcome, even after making remarkable progress with respect to the provision of rural water supply services in India, many habitations and villages continue to face water scarcity during the summer months. The problem is more pronounced in areas underlain by hard rocks and where wells are the source of water supply. As discussed in Chaps. 2 and 3, in such areas, due to the constraints imposed by the geo-hydrological settings (with no primary porosity in rocks and the shallow formations), the utilizable groundwater recharge is very extremely limited.

Taking cognizance of the lack of scientific assessments concerning the effectiveness of the water supply schemes with respect to institutional and technological models adopted as per the policy reforms, an attempt is made to compare the performance of selected rural water supply schemes (RWSS) in different geological settings of Maharashtra and find out the model of water supply technology which works best under each such setting. For this purpose, functional water supply schemes from different divisions of Maharashtra were shortlisted and compared in terms of their physical performance (water supply coverage, water supply levels and use, quality of supplied water, access to water, duration and frequency of water supply, reliability), economic performance (cost of water supply, capital, and O&M), and financial performance (rate of recovery of water charges). Also, the impact of reliable water supply on the adoption of improved toilets was assessed.

4.2 Review of Studies on Effectiveness of Rural Water Supply Schemes

The World Bank (2008) undertook an analysis of water supply schemes in 10 states that accounted for about 60% of India's rural population. Using a large body of data on various aspects of rural water supply in ten states (Andhra Pradesh, Karnataka, Kerala, Maharashtra, Orissa, Punjab, Tamil Nadu, Uttar Pradesh, Uttarakhand, and West Bengal), the effectiveness of schemes is analysed across technology (handpumps, mini water schemes, single-village schemes, multi-village, and regional schemes) and institutional arrangements: schemes managed by state utilities (engineering departments, boards, and so on), district government functionaries (Zilla Parishad), village local government (Gram Panchayat), and communities. Decentralization and reform receive particular attention in the study. A comparative assessment of performance is made of schemes under the 'demand responsive' reform programmes and those under the traditional 'supply' (target)-driven mode. The conclusions of the study are as follows:

1. Rural water supply schemes are commonly weak in performance. The quantity of water supply and hours of supply commonly fall short of design, especially in summer. Sizeable sections of the households face problems caused by frequent breakdowns, non-availability of daily supply, and insufficient supply compared to the requirement. Further, the quality of water supplied fails to meet standards in a sizeable proportion of cases.
2. Rural households are bearing huge coping costs. The most important component of coping cost is the opportunity cost of the time spent on water collection. Other components include the expenditure incurred by households on the repair of public water sources and for the maintenance of household equipment for private water supply arrangements.
3. Supply-driven programmes incur large institutional costs, substantially raising the cost of service-provision. The average level of institutional cost in supply-driven programmes (24%) far exceeds that in demand-driven programmes (11%).
4. Capital cost of piped water supply schemes is excessive. There are indications that the overall cost of rural water supply infrastructure can appreciably be reduced by increasingly shifting to a demand-driven mode. The cost of the infrastructure of piped water supply schemes implemented by the community is far less than the cost of schemes implemented by public utilities or government agencies.
5. O&M expenditure is inadequate, causing schemes to perform below design and shortening their useful life. On average, the actual O&M expenditure of piped water schemes is about half of the good practice 'design performance O&M cost'.
6. Significant wastage of resources arises from over-provisioning by some schemes, defunct schemes, and the existence of multiple schemes.
7. Low O&M cost recovery: Overall, the O&M cost recovery in piped water schemes is about 46% (in handpump schemes, cost recovery is almost nil). Owing to the low recovery of cost, there is a huge subsidy to rural water supply, constituting about 0.2–0.4% of the SDP of the states studied. On an average, the subsidy is 160% of state rural water supply funds. The cost recovery performance of community-managed schemes (71% on average) is distinctly better than the public utility-managed schemes (21%). The Gram Panchayat (GP)-managed schemes have higher cost recovery than government/public utility-managed schemes, but less than that of community-managed schemes.
8. Effectiveness of schemes is 'low to moderate'. Seventeen important indicators were developed for piped water supply schemes, and four indices of effectiveness were constructed, representing: (a) adequacy and reliability, (b) affordability, (c) environmental sustainability, and (d) financial sustainability. The worst performance is found in respect of financial sustainability. In regard to the effectiveness of piped water supply schemes, no great differences are found

between technologies or between institutional arrangements for the management of schemes. However, community-managed schemes are found to be doing somewhat better than public utility-managed schemes.

9. Only a part of the public expenditure reaches the beneficiaries. The leakages are relatively greater for supply-driven programmes than for demand-driven programmes. Out of the Rs. 100 spent, while Rs. 75 trickles down to the beneficiaries in the 'demand-driven programmes', only about Rs. 50 reaches the beneficiaries in 'supply-driven programmes'.
10. Multi-village schemes (including regional schemes) cost much more than single-village schemes.

MacAllister et al. (2020) analysed the performance of a wide range of water source types in Ethiopia, using a unique dataset of more than 5000 individual water points. The monitoring survey and analysis focussed on five main water source types: boreholes equipped with handpumps; motorized boreholes; protected springs and springs providing gravity-fed supplies; protected hand-dug-wells and open sources (such as unprotected wells and surface water). These water source types represent the most common technologies used to provide water to rural communities in Ethiopia and across much of sub-Saharan Africa. The study found that mean functionality ranged from 60% for motorized boreholes to 75% for hand-pumped boreholes. Real-time monitoring and responsive operation and maintenance led to rapid increases in the functionality of hand-pumped and, to a lesser extent, motorized boreholes. Increased demand was placed on motorized boreholes in lowland areas as springs, hand-dug-wells, and open sources failed. Boreholes accessing deep (>30 m) groundwater performed best during the drought. Prioritizing access to groundwater via multiple improved sources and a portfolio of technologies, such as hand-pumped and motorized boreholes, supported by responsive and proactive operation and maintenance, increases rural water supply resilience.

Wilbers et al. (2014) conducted a study in the Mekong Delta (MD) in Vietnam, where piped water supply stations are being intensively built to provide safe and clean drinking water resources to communities. A total of 542 households were interviewed in the rural areas. The interviews aimed to assess the availability, usage, and households' perceptions of piped water supplies. The study found that in total, 39% of households interviewed had potential access to piped water distribution systems. In contrast, the other households (61%) had no possible access to piped water, since a water supply station was not present or was not operational. More than half of the interviewed households with access to a piped water supply did not use this supply as a source of drinking water due to (i) high connection fees; (ii) preference for other water sources; and (iii) perceived poor quality/quantity.

The reliability of supplied piped water was another concern in some of the studied areas, which led to the fact that households used more reliable sources like groundwater and even surface water. The study also showed that the maintenance and distribution of water supply stations should significantly improve for piped water to become a reliable drinking water source. Additionally, alternatives, such as rainwater harvesting and decentralized treatment facilities, should also be considered.

Overall, the review suggests that the available evidence is on the performance of rural water supply schemes in general vis-à-vis attributes such as adequacy and reliability of supplies, cost levels, O&M expenditure, O&M cost recovery and sustainability, and does not correspond to the comparative performance of schemes that use different types of sources. The study by World Bank (2008) involved comparative analyses of schemes managed under the traditional 'supply-driven approach' and under 'demand-driven approach' by the communities vis-à-vis the capital and operating costs and found notable differences. The bank study, however, did not find any major difference in the effectiveness of piped water supply schemes with changes in water supply technology and institutional arrangements for management. The study in Ethiopia compared the effectiveness of schemes that use different technologies and types of sources, such as motorized bore wells, handpumps, open wells, springs, and unprotected surface water sources. The study in the Mekong River delta of Vietnam looked at the access to piped water, the extent of use of water by the households which had access, and the community's perceptions of piped water supply vis-à-vis the cost and water quality. However, there is no study that actually compares the schemes by 'source water', such as managed surface reservoirs and groundwater-based sources for effectiveness.

4.3 Selection of Water Supply Schemes for Performance Assessment

The major criteria for the selection of rural water supply schemes were that they should represent the different geological and hydrological settings, resource base tapped, and technologies used for rural water supply delivery in the state.

The six divisions represent the different geological and hydrological set up in Maharashtra. The Konkan division is a coast mostly having laterite and alluvium aquifers and receives the highest rainfall (long term average annual rainfall is 2988 mm) in the state, Nagpur division has a mix of basalt, gneiss, and intrusives aquifers and gets an average annual rainfall of 1500 mm, Aurangabad and Pune divisions have basaltic aquifers and receive the lowest rainfall in the state (long-term average annual rainfall of about 750 mm), and Nashik and Amravati divisions have basalt aquifers with alluvium in the northern parts along the Tapi river and an average annual rainfall of 880 mm.

Further, while selecting the schemes, care was also taken to ensure that two different types of schemes from the same type of geological and hydrological setting were selected for comparison. This is to nullify the influence of the physical environment on the performance of schemes.

Considering all the criteria, overall 12 schemes, two each from the representative district of every division in Maharashtra were selected for assessing the performance. The districts covered include Chandrapur, Amravati, Palghar, Nandurbar, Satara, and Nanded. The selected schemes were both single village and multi-village. The types of sources for these schemes were surface water and groundwater. Different technologies covered were bore well, dug well, handpump, river lifting, and reservoir.

The primary data for assessing the performance of the schemes were collected from representative households served by the selected 12 schemes using a questionnaire and a recall method. The fieldwork for primary data collection was conducted from September to November 2020. The data were analysed using simple statistical methods such as sample size distribution, mean, etc.

4.4 Scheme Characteristics and Respondents Details

The details on scheme type and its coverage are presented in Table 4.1. Out of the 12 selected schemes in total, three schemes were based on dug well, two on bore well, one on handpump, four on surface water reservoirs, and two on river lifting.

Table 4.1 Details of the selected rural water supply schemes in Maharashtra for the survey

District covered (division)	Scheme type	Villages covered	Scheme started	Coverage (number of)			% HH connections
				Villages	Habitations	HHs	
Chandrapur (Nagpur)	Dug well PWS	Single	2010	1	1	180	67
	River lift PWS	Multiple	2012	3	3	352	26
Amravati (Amravati)	Dug well PWS	Single	2012	1	1	158	56
	River lift PWS	Single	2011	1	1	142	73
Palghar (Konkan)	Bore well PWS	Two	2019	2	9	883	48
	Reservoir PWS	Multiple	2004	10	22	6212	11
Nandurbar (Nashik)	Handpump	Single	2017	1	1	432	0
	Reservoir PWS	Single	2016	1	1	318	82
Satara (Pune)	Dug well PWS	Single	1970	1	2	955	100
	Reservoir PWS	Multiple	2001	7	27	1231	14
Nanded (Aurangabad)	Bore well PWS	Single	2010	1	1	670	0
	Reservoir PWS	Single	2004	1	1	630	0

Source Authors' analysis of the field data

Except for the one handpump-based scheme, all schemes had a provision of piped water supply (PWS). While most of the selected groundwater-based schemes serve only a village, the surface water-based schemes cater to several villages. However, groundwater-based schemes have a high proportion of household (HH) connections than the surface water schemes, except for the schemes in Amravati and Nandurbar. Further, HHs served by the groundwater-based scheme in Nandurbar and both the selected schemes in Nanded had no individual water supply connection.

For the two selected schemes from each division, a minimum of 40 HHs were surveyed on aspects related to quantity and quality of water supplied for various uses, frequency and reliability of supply, access to water supply, and the overall status of the water supply. Further, data on the scheme-wise planned and actual water supply, duration of water supply, and water tax recovery rate were collected from the concerned Village Water and Sanitation Committee (VWSC) and/or Gram Panchayat (GP). Table 4.2 provides the details of the HHs surveyed for each scheme. HHs reported accessing water from the schemes from as recent as two years to as long as 50 years.

Table 4.2 Details of the rural HHs surveyed

District (division)	Primary scheme	Block surveyed	GP or village surveyed	HH surveyed	Average family size	Average no. of years for which scheme is used
Chandrapur (Nagpur)	Dug well	Rajura	Kalamana	19	4.2	8
	River lift	Korpana	Bharosa	22	4.7	8
Amravati (Amravati)	Dug well	Dharani	Jhyapal	23	4.2	8
	River lift	Dharani	Tembli	23	3.7	5
Palghar (Konkan)	Bore well	Dahanu	Dabhadi	21	8.0	2
	Reservoir	Jawhar	Kortad	24	5.0	15
Nandurbar (Nashik)	Handpump	Akkalkuwa	Amli	22	5.5	3
	Reservoir	Nandurbar	Nimbhel	20	5.4	3
Satara (Pune)	Dug well	Khatav	Diskal	21	6.3	50
	Reservoir	Phaltan	Saskal	20	5.8	20
Nanded (Aurangabad)	Bore well	Mukhed	Eklara	20	5.7	10
	Reservoir	Deglur	Khanapur	20	3.6	10

Source Authors' analysis of the field data

4.5 Scheme Performance

4.5.1 Water Supply and Use

The planned water supply from the selected schemes varies from 40 to 50 lpcd (litres per capita per day) for the groundwater-based schemes and 50–55 lpcd for the schemes based on surface water (please refer to Table 4.3). In terms of actual water supply across seasons (monsoon, winter, and summer), surface water schemes fare better than their groundwater counterparts. Except for Nanded, the surface water schemes supplied water as planned, the highest reported supply (55 lpcd) being in Amravati and lowest in Nanded (27 lpcd). The groundwater-based schemes in Palghar and Nandurbar were among those having the lowest water supply, with 13 and 16 lpcd, respectively, in the monsoon that further reduces to 9 lpcd during summers in the case of Palghar.

Apart from the selected water supply scheme (i.e. the primary source), HHs reported dependence on other sources of water as well. The number of water sources (often unimproved ones) was high for low-priority uses that do not require high-quality water, such as other domestic uses, sanitation, and livestock use (please refer to Table 4.4). For instance, all the households (except for those in Amravati and Nandurbar) depend on the only source of water (i.e. the primary scheme) for meeting drinking and cooking requirements. For other domestic uses, which include washing utensils, cleaning the house, bathing, sanitation, etc., most of the households were

Table 4.3 Scheme-wise planned and actual average water supply

District (division)	Primary scheme	Planned (lpcd)	Actual water supplied (lpcd)		
			Monsoon	Winter	Summer
Chandrapur (Nagpur)	Dug well PWS	50	35	35	35
	River lift PWS	–	–	–	–
Amravati (Amravati)	Dug well PWS	–	–	–	–
	River lift PWS	55	55	55	55
Palghar (Konkan)	Bore well PWS	40	13	11	9
	Reservoir PWS	–	34	34	34
Nandurbar (Nashik)	Handpump	50	16	16	16
	Reservoir PWS	50	48	48	48
Satara (Pune)	Dug well PWS	–	–	–	–
	Reservoir PWS	55	49	49	41
Nanded (Aurangabad)	Bore well PWS	–	–	–	–
	Reservoir PWS	50	27	27	27

Source Authors' analysis of the field data
Note Data could not be obtained on certain parameters for some schemes and hence not reported

Table 4.4 Number of water supply sources that the surveyed HHs access for meeting various needs

District (division)	Primary scheme	Total number of water supply sources accessed			
		Drinking and cooking	Other domestic	Sanitation	Livestock use
Chandrapur (Nagpur)	Dug well PWS	1	2	2	2
	River lift PWS	1	2	1	1
Amravati (Amravati)	Dug well PWS	2	2	2	No livestock
	River lift PWS	2	2	2	No livestock
Palghar (Konkan)	Bore well PWS	1	2	2	3
	Reservoir PWS	1	1	1	1
Nandurbar (Nashik)	Handpump	2	2	2	2
	Reservoir PWS	2	2	2	3
Satara (Pune)	Dug well PWS	1	1	1	No livestock
	Reservoir PWS	1	1	1	No livestock
Nanded (Aurangabad)	Bore well PWS	1	1	1	1
	Reservoir PWS	1	1	1	1

Source Authors' analysis of the field data

dependent on two sources. For Palghar (bore well scheme) and Nandurbar (reservoir scheme), the number of dependent sources increases to three for meeting the water requirement of livestock. In both these districts, the water supplied through the primary water supply scheme (groundwater-based) is among the lowest.

Nevertheless, households served by the reservoir scheme in Palghar and those served by both the surface water and groundwater schemes in Satara and Nanded depend on only one source.

Further, except for Nanded, the per capita water use (from all the sources) was reported to be higher for households that are dependent on surface water schemes than those on groundwater schemes (please refer to Table 4.5). This is largely due to the higher per capita supply from the surface water schemes (as presented in Table 4.3). Among the groundwater-based schemes, households dependent on the dug well-based scheme in Chandrapur reported the lowest average per capita water use.

Table 4.5 Average per capita water use for drinking, domestic, and sanitation from all the sources

District (division)	Primary scheme	Drinking and cooking (lpcd)[a]			Other domestic (lpcd)			Sanitation (lpcd)		
		M	W	S	M	W	S	M	W	S
Chandrapur (Nagpur)	Dug well PWS	5.9	6.7	7.7	5.9	6.5	7.3	10.6	12.6	14.6
	River lift PWS	9.3	9.1	10.8	26.4	28.2	32.0	10.5	10.8	12.4
Amravati (Amravati)	Dug well PWS	22.9	22.9	27.0	28.0	28.0	34.1	9.4	9.4	10.6
	River lift PWS	25.6	25.6	28.4	30.2	30.2	35.1	7.0	7.0	7.0
Palghar (Konkan)	Bore well PWS	9.4	9.0	11.0	13.9	13.8	15.9	8.5	8.4	9.5
	Reservoir PWS	8.3	8.1	10.4	22.4	21.3	25.9	10.5	10.6	12.5
Nandurbar (Nashik)	Handpump	10.3	10.3	10.3	26.9	26.5	26.5	13.4	13.4	13.4
	Reservoir PWS	9.0	9.0	9.1	29.9	30.0	30.1	17.5	17.5	17.5
Satara (Pune)	Dug well PWS	5.9	5.9	7.9	30.7	22.5	32.8	18.0	18.0	16.0
	Reservoir PWS	5.9	5.9	7.9	30.6	22.4	32.8	18.0	18.0	16.0
Nanded (Aurangabad)	Bore well PWS	6.3	6.3	6.6	26.1	26.1	34.1	21.0	21.0	20.9
	Reservoir PWS	10.7	10.7	10.7	10.7	10.7	10.7	10.7	10.7	10.7

Source Authors' analysis of the field data
Note [a]M represents monsoon months, W winter months, and S summer months

Except for Nanded, most of the households reported good quality of the water supplied by the primary scheme (refer to Table 4.6). For the surveyed households in Nanded, not only is the water supplied by the surface water scheme the lowest among all such schemes, but the quality of water is also not good.

4.5.2 *Access to Water Supply*

Overall, a large proportion of the surveyed HHs depends on surface water schemes that have a tap either inside the dwelling or in the dwelling premises (please refer to Table 4.7). In Chandrapur, 100% of the surveyed HHs dependent on surface water source tap water within the dwelling premises, and in Satara, all the surveyed HHs

Table 4.6 Percentage of surveyed HHs reporting good quality water supply

District (division)	Primary scheme	% HHs reported good quality water			
		Drinking and cooking	Other domestic	Sanitation	Livestock use
Chandrapur (Nagpur)	Dug well PWS	100	100	100	100
	River lift PWS	100	100	100	100
Amravati (Amravati)	Dug well PWS	96	96	96	No livestock
	River lift PWS	96	96	96	No livestock
Palghar (Konkan)	Bore well PWS	100	100	86	100
	Reservoir PWS	100	100	88	100
Nandurbar (Nashik)	Handpump	100	100	91	100
	Reservoir PWS	100	100	90	100
Satara (Pune)	Dug well PWS	100	100	100	No livestock
	Reservoir PWS	100	100	100	No livestock
Nanded (Aurangabad)	Bore well PWS	0	0	0	0
	Reservoir PWS	0	0	0	0

Source Authors' analysis of the field data

were found to be getting water from the surface water scheme that have taps inside the dwelling.

However, in the case of the groundwater-based scheme, most of the surveyed households access water from sources away from their dwelling. On average, they reported travelling between 65 and 307 m to access water (Fig. 4.1), except in Chandrapur and Satara where they have a connection either inside the dwelling or within the dwelling premise (please refer to Table 4.7). Nevertheless, most of the surveyed HHs, whether dependent on surface water or groundwater, reported improvement in access to water under the running scheme in comparison with the old scheme.

In terms of supply duration, most of the households reported fewer supply hours than planned by the water supply agency (refer to Table 4.8). Overall, the water supply hours reported by the HHs vary from 0.5 to 3.0 h/day for groundwater schemes and 0.5 to 2.0 h/day for surface water schemes. Though the HHs connected to groundwater schemes reported a higher duration of water supply, the total quantity of water supplied was less than that of surface water schemes (refer to Table 4.3).

Table 4.7 HH's access to water supply (primary scheme)

District (division)	Primary scheme	% HH reported (current scheme)		
		Tap within dwelling	Tap within HH premises	Stand post
Chandrapur (Nagpur)	Dug well PWS	–	100	–
	River lift PWS	–	100	–
Amravati (Amravati)	Dug well PWS	35	13	52
	River lift PWS	96	–	4
Palghar (Konkan)	Bore well PWS	–	48	52
	Reservoir PWS	25	67	8
Nandurbar (Nashik)	Handpump	–	23	77
	Reservoir PWS	–	95	5
Satara (Pune)	Dug well PWS	100	–	–
	Reservoir PWS	100	–	–
Nanded (Aurangabad)	Bore well PWS	–	–	100
	Reservoir PWS	15	85	–

Source Authors' analysis of the field data

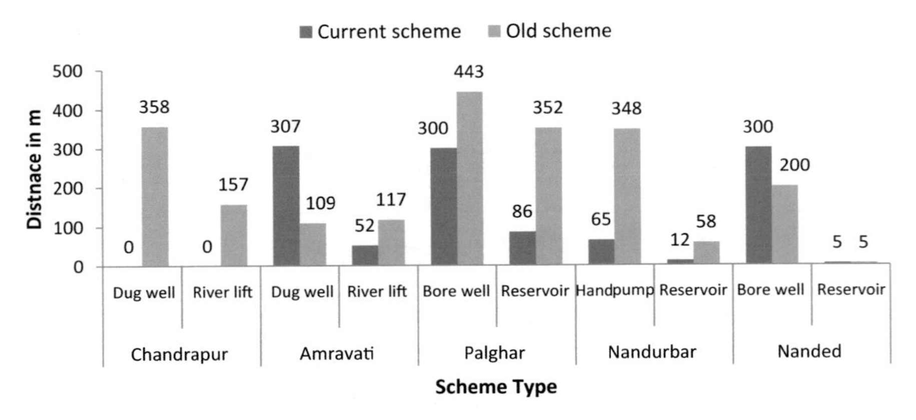

Fig. 4.1 Average distance travelled by surveyed households to access water. *Source* Authors' analysis of the field data

4.5.3 *Frequency of Water Supply*

Overall, there is an improvement in the frequency of water supply as most of the HHs served by the existing water scheme receive water daily in comparison with the old water supply scheme (please refer to Table 4.9). Further, most of the households (irrespective of the scheme) receive water during the morning hours. However, all

Table 4.8 Water supply hours for the primary source

District (division)	Primary scheme	Reported by the water agency (h/day)			Reported by the HHs (h/day)		
		M	W	S	M	W	S
Chandrapur (Nagpur)	Dug well PWS	3.0	3.0	3.0	1.9	1.9	1.9
	River lift PWS	2.0	2.0	2.0	2.0	2.0	2.0
Amravati (Amravati)	Dug well PWS	2.0	2.0	2.0	2.0	2.0	2.0
	River lift PWS	2.0	2.0	2.0	1.0	1.0	1.0
Palghar (Konkan)	Bore well PWS	4.0	4.0	4.0	1.0	1.0	0.9
	Reservoir PWS	2.5	2.5	2.5	0.5	0.5	0.5
Nandurbar (Nashik)	Handpump	4.0	4.0	4.0	2.0	2.0	2.0
	Reservoir PWS	4.0	4.0	4.0	0.6	0.6	0.6
Satara (Pune)	Dug well PWS	0.5	0.5	0.5	0.5	0.5	0.3
	Reservoir PWS	1.0	1.0	1.0	1.0	1.0	1.0
Nanded (Aurangabad)	Bore well PWS	3.0	3.0	4.0	3.0	3.0	5.0
	Reservoir PWS	3.0	3.0	3.0	2.0	2.0	2.0

Source Authors' analysis of the field data

Table 4.9 Percentage of surveyed HHs reporting daily supply (frequency) of water

District (division)	Existing primary scheme	% HH reported daily supply (old scheme)			% HH reported daily supply (existing primary scheme)		
		M	W	S	M	W	S
Chandrapur (Nagpur)	Dug well PWS	100	100	84	100	100	89
	River lift PWS	100	100	100	100	100	100
Amravati (Amravati)	Dug well PWS	100	100	100	100	100	100
	River lift PWS	100	100	100	0	0	0
Palghar (Konkan)	Bore well PWS	90	90	90	100	100	100
	Reservoir PWS	100	100	100	100	100	100
Nandurbar (Nashik)	Handpump	100	100	100	100	100	100
	Reservoir PWS	100	100	60	100	100	70
Satara (Pune)	Dug well PWS	–	–	–	0	0	0
	Reservoir PWS	–	–	–	0	0	0
Nanded (Aurangabad)	Bore well PWS	100	100	100	100	100	100
	Reservoir PWS	100	100	100	100	100	100

Source Authors' analysis of the field data
Note Data could not be obtained on certain parameters for some schemes and hence not reported

the surveyed HHs dependent on the surface water schemes in Amravati and Satara and groundwater scheme in Satara receive water on alternate days.

Further, most of the surveyed households in Chandrapur, Amravati, Satara, and Nanded can meet most of their water requirements under the current water supply scheme. However, for Palghar and Nandurbar, though there is an improvement in comparison with the previous scheme, a substantial proportion of HHs were unable to meet all their water requirements from the current primary source (please refer to Table 4.10). This is also reflected in the higher number of sources that HHs in Palghar and Nandurbar access for meeting their water requirement (please refer to Table 4.4).

While the option of depending on alternate sources of water that either yield water of inferior quality near the dwelling or yield water of good quality but are located away from the dwelling, is not available, rural households also depend on tanker water supply as our secondary data analysis has shown. Such a situation generally emerges during the summer months. Such instances were noticed in three of the six locations, viz. Chandrapur, Satara, and Nanded. The source was a private tanker in all three cases. The proportion of the sample households depending on tanker water supply during summer months was a lowest of 15% in Nanded to a highest of 54% in Satara. The average quantum of water purchased ranged from 2.4 kl (kilolitres) in the case of the scheme in Nanded to 3.64 kl in the case of the scheme in Satara. The prices were also higher in the case of Satara (INR[1] 0.35/l). Please refer to Table 4.11 for details.

One reason for a relatively lower incidence of tanker water supply could be due to the purposive sampling of the schemes, wherein we are only choosing schemes that are currently functional. The other reason could be the small sample size of 12 schemes, which is surely not representative of the condition vis-à-vis water supply in the state.

4.5.4 *Maintenance of Water Supply Scheme and Water Tax Recovery*

As per the information from the surveyed HHs, most of the water supply schemes breakdown once every three or six months and need repair. While the average time taken for repair of surface water schemes is usually 2–3 days, the groundwater schemes usually take longer (please refer to Table 4.12). For instance, in the case of the groundwater scheme in Chandrapur and Nandurbar, a majority of the households reported that repair took about a week.

Water tax collection is an indication of the scheme's reliability, i.e. of obtaining water of adequate quality in adequate quantity, with reasonable frequency and duration. Generally, people are willing to pay for good services. In the case of the groundwater schemes, the water tax collection rate varied from 30 to 79%, whereas for

[1] As of August 2021, 1 USD equals to INR 74.

Table 4.10 Percentage of HH water requirement met by the previous and current primary water supply scheme

District (division)	Existing primary scheme	Drinking and cooking		Washing and toilet use		House cleaning		Livestock use	
		Previous	*Current*	*Previous*	*Current*	*Previous*	*Current*	*Previous*	*Current*
Chandrapur (Nagpur)	Dug well PWS	100	100	65	100	48	100	100	98
	River lift PWS	100	100	67	100	71	98	100	100
Amravati (Amravati)	Dug well PWS	90	100	90	100	90	100	No livestock	
	River lift PWS	70	98	71	98	71	98	No livestock	
Palghar (Konkan)	Bore well PWS	74	89	60	75	51	65	47	65
	Reservoir PWS	72	92	60	78	56	74	54	68
Nandurbar (Nashik)	Handpump	82	91	77	93	57	78	79	88
	Reservoir PWS	80	88	83	89	78	85	79	91
Satara (Pune)	Dug well PWS	–	100	–	100	–	100	No livestock	
	Reservoir PWS	–	100	–	100	–	100	No livestock	
Nanded (Aurangabad)	Bore well PWS	100	100	70	100	80	100	100	100
	Reservoir PWS	100	100	100	100	100	100	100	100

Source Authors' analysis of the field data
Note Data could not be obtained on certain parameters for some schemes and hence not reported

Table 4.11 Details on water purchase by the households surveyed (HHs only under these three schemes reported water purchase)

District	Chandrapur	Satara	Nanded
Scheme surveyed	Groundwater-based		
Type	Bore well PWS		
Coverage	Single GP		
HH surveyed	19	21	20
Average family size	4	6	6
No. (%) of HHs purchasing water	4 (21%)	11 (54%)	3 (15%)
Source	Private tanker		
Avg. amount of water purchased during the entire summer season (l/HH)[a]	2940	3636	2400
Avg. cost of purchased water (Rs./l)	0.24	0.35	0.23
Avg. amount spent per HH during the entire summer season (Rs.)	698	1273	540

Source Authors' analysis of the field data
Note [a]Considering only the households that are engaged in tanker water purchase

Table 4.12 Time taken for repairing a water supply scheme

District (division)	Primary scheme	% HHs reported		
		Within a day	Within 2–3 days	Within a week
Chandrapur (Nagpur)	Dug well PWS	–	11	89
	River lift PWS	–	55	45
Amravati (Amravati)	Dug well PWS	–	100	–
	River lift PWS	–	100	–
Palghar (Konkan)	Bore well PWS	10	76	14
	Reservoir PWS	13	54	33
Nandurbar (Nashik)	Handpump	–	9	64
	Reservoir PWS	35	45	–
Satara (Pune)	Dug well PWS	–	100	–
	Reservoir PWS	100	–	–
Nanded (Aurangabad)	Bore well PWS	–	100	–
	Reservoir PWS	–	100	–

Source Authors' analysis of the field data

surface water schemes, it varied from 19 to 94% (please refer to Fig. 4.2). Apart from one scheme (in Nandurbar), the recovery rate is high and better for the surface water schemes, compared to groundwater-based schemes.

Further, a significantly higher proportion of HHs served by surface water schemes reported improvement in the overall reliability of the water supply, whereas for

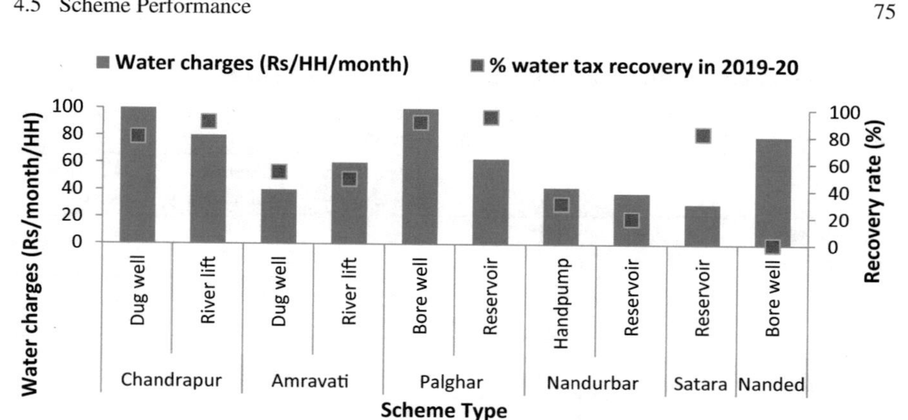

Fig. 4.2 Percentage water tax collection for the selected water supply schemes. *Source* Authors' analysis of the field data. *Note* Data could not be obtained on certain parameters for some schemes and hence not reported

groundwater schemes, the improvement was either in the adequate quantity or in the physical access to water (please refer to Table 4.13). Thus, it can be inferred that the surface water schemes are more reliable than groundwater-based schemes.

4.5.5 Adoption of Improved Toilet

In many rural areas, the built-up toilets are not used due to a shortage of water. The research studies have shown that improved water supply influences the HH decision to adopt and use a toilet as water is no longer a constraining factor. An improved toilet is more likely to prevent human contact with human excreta and thus offer greater health benefits in comparison with the unimproved toilets (WHO and UNICEF 2006). Examples of improved toilets include flush or pour-flush to a piped sewer system or septic tank, ventilated improved pit latrine, pit latrine with slab, and composting toilet. In rural areas of India, households prefer pour-flush to septic tanks and pit latrines with slabs, provided that a reliable water supply is available.

Except for Nanded and Nandurbar, there is not much difference in the adoption of improved toilets by the households served by groundwater and surface water. Further, the overall toilet adoption by the HHs under both the schemes in Amravati and the groundwater-based scheme in Nanded is among the lowest (please refer to Fig. 4.3). For the latter, the adoption of toilets is low as no household served by the groundwater scheme has an individual tap connection.

Table 4.13 Percentage HHs reported an overall improvement in the water supply

District (division)	Primary scheme	% HHs reported improvement	In terms of (% of the HHs who reported improvement)		
			Adequate quantity	Reliability of water supply	Physical access
Chandrapur (Nagpur)	Dug well PWS	100	26	74	–
	River lift PWS	100	73	23	5
Amravati (Amravati)	Dug well PWS	100	96	–	–
	River lift PWS	100	91	–	9
Palghar (Konkan)	Bore well PWS	95	5	10	85
	Reservoir PWS	92	32	–	68
Nandurbar (Nashik)	Handpump	100	–	–	100
	Reservoir PWS	95	–	95	5
Satara (Pune)	Dug well PWS	100	43	48	10
	Reservoir PWS	100	30	70	–
Nanded (Aurangabad)	Bore well PWS	100	–	100	–
	Reservoir PWS	5	–	100	–

Source Authors' analysis of the field data

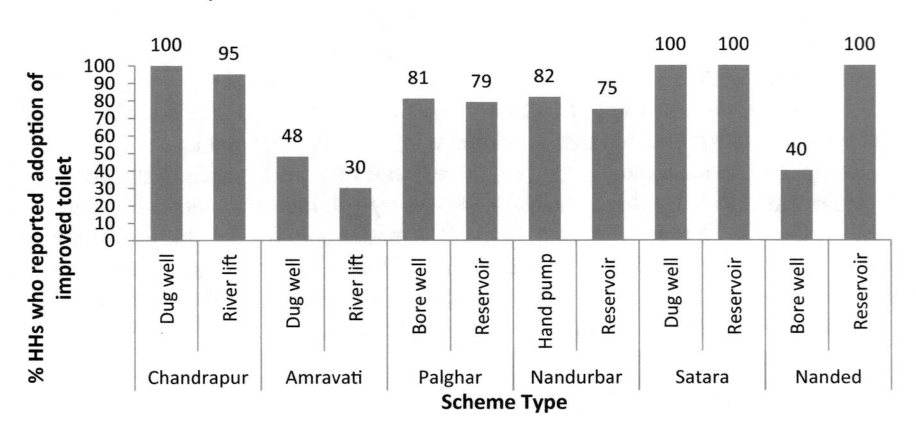

Fig. 4.3 Proportion of HHs reported adoption of improved toilet. *Source* Authors' analysis of the field data

4.6 Findings

The following inferences can be drawn from the analysis of field survey data on the performance of the rural water supply schemes:

- Surface water-based rural water supply schemes show better performance in terms of the actual quantity of water supply and access to the water supply.
- The average per capita supply for surface water-based schemes was higher than that of groundwater schemes. For surface water schemes, the daily per capita supply was highest for a scheme in Amravati (55 lpcd) and lowest in Nanded (27 lpcd). For the groundwater-based schemes, the average daily water supply was highest for Chandrapur (35 lpcd) and lowest in Palghar (13 lpcd) and Nandurbar (16 lpcd).
- Overall, HHs served by the reservoir-based schemes depend on only the primary source to meet the household water needs, whereas those served by groundwater-based schemes depend on at least two sources of water. In Palghar and Nandurbar, the number of dependent sources increase to three for meeting the water requirement of livestock as the water supplied through the primary water source (groundwater-based) was among the lowest.
- Except for Nanded, the per capita water use (from all the sources) was reported to be higher for households that are dependent on surface water-based schemes than those dependent on groundwater-based schemes. Nanded is in the Marathwada region that receives the lowest rainfall in Maharashtra.
- Among the groundwater-based schemes, households dependent on the dug well-based scheme in Chandrapur reported the lowest water use. Chandrapur is in the Vidarbha region that experiences seasonal groundwater scarcity, and hence, very little water is normally available in wells during summer.
- Overall, a large proportion of the surveyed HHs dependent on surface water-based schemes either has a water tap inside the dwelling or within the dwelling premise. However, in the case of the groundwater schemes, most of the surveyed households' access water outside their dwelling or away from their dwelling premise. On average, they reported travelling between 65 and 307 m to access water from the public source.
- Though the HHs connected to groundwater-based schemes reported a higher duration of water supply—0.5–3.0 h/day for groundwater schemes to 0.5–2.0 h/day for surface water schemes—the total quantity of water supplied was less than that of surface water schemes.
- Though most of the households (irrespective of the scheme) receive water daily during the morning hours, the entire surveyed HHs dependent on the surface water scheme in Amravati and Satara and those on the groundwater scheme in Satara received water on alternate days.
- Considering both types of schemes, most of the surveyed households in Chandrapur, Amravati, Satara, and Nanded could meet most of their requirements under the current scheme. However, for Palghar and Nandurbar, though the current scheme shows better performance than the old one, a substantial proportion of

HHs were unable to meet all their water requirements from the current source. It is quite likely that the household tap water from alternative sources may meet their water needs. It must be mentioned here that Palghar is in the Konkan division that receives the highest rainfall in Maharashtra and Nandurbar is in the Nashik division that has alluvial tracts; therefore, availability of alternate surface water and groundwater is not a constraint.

- The instance of households depending on (private) tanker water supply during summer, a clear indication of the failure of the public schemes, was encountered in three out of the six cases, i.e. in Nanded, Chandrapur, and Satara. The proportion of households depending on tanker water supply varied from 15% in Nanded to 54% in Satara. The amount of water purchased by these HHs ranged from 2.4 kl per year in Nanded to 3.64 kl per year in Satara and the price of water ranged from INR 0.23/l in Nanded to INR 0.35/l in Satara.
- While the average time taken for repair of surface water-based schemes is usually 2–3 days, the groundwater schemes usually take longer. For instance, in the case of the groundwater scheme in Chandrapur and Nandurbar, repair takes about a week. This is because of the difference in management structures between the two types of schemes.
- Apart from one scheme (in Nandurbar), water tax collection is high and slightly better for the surface water schemes—19–94% against 30–79% for groundwater water-based schemes. Also, the adoption of improved toilets is more for HHs receiving water from surface water schemes as a large proportion of the surveyed HHs have a tap connection inside the dwelling or within the dwelling premises.

4.7 Conclusions and Policy Inferences

A review of the limited studies looking at the performance of rural water supply schemes worldwide suggests that the available evidence is on the performance of schemes in general vis-à-vis attributes such as adequacy and reliability of supplies, cost levels, O&M expenditure, O&M cost recovery and sustainability, and they do not correspond to the comparative performance of schemes that use different types of sources. The study by the World Bank (2008) conducted in India involved comparative analyses of schemes managed under the traditional 'supply-driven approach' and under 'demand-driven approach' by the communities vis-à-vis the capital and operating costs and found notable differences. The study in Ethiopia compared the effectiveness of schemes that use different technologies and different types of sources, such as motorized bore wells, handpumps, open wells, springs, and unprotected surface water sources. The study in the Mekong Delta in Vietnam, where piped water supply schemes were built to provide safe and clean drinking water to communities, assessed the access to water supply connections, usage of water, and households' perceptions of piped water supplies. However, there is no study that actually compares the schemes by 'source water', such as managed surface reservoirs and groundwater-based sources for effectiveness.

In this study conducted in Maharashtra, an attempt was made to compare the performance of selected rural water supply schemes (RWSS) in different geological settings and find out the model of water supply technology that works better under each such setting. For this purpose, functional water supply schemes from different divisions of Maharashtra were shortlisted and compared in terms of their physical performance (water supply coverage, water supply levels and use, quality of supplied water, access to water, duration and frequency of water supply, reliability), economic performance (cost of water supply, capital and O&M), and financial performance (rate of recovery of water charges). Also, the influence of the reliability of water supply on the adoption of improved toilets was assessed.

The study found the overall performance of water supply schemes based on surface reservoirs and direct lifting from rivers to be better than that of schemes based on groundwater, when indicators such as water supply coverage, adequacy of water supply, physical access to water supply source, quality of the supplied water, annual maintenance, and revenue recovery are considered for comparison. Further, in regions with similar physical characteristics, vis-à-vis hydrology, geohydrology, and topography, the regional water supply schemes and single-village schemes that are based on surface reservoirs ensured water supplies of higher dependability to the rural areas, as compared to schemes that are dependent on wells.

In hard rock areas having intensive agriculture with no surface irrigation facilities and heavy dependence of farmers on wells for meeting crop water needs, groundwater-based schemes are unable to provide sufficient quantities of water for meeting domestic water requirements throughout the year, and scheme failure during summer months due to drying up of wells and yield reduction is very common. This is a major aspect that needs to be factored in the investment decision-making framework for building water supply schemes, as scheme failure inflicts a hidden cost on the communities and the utility. While the water supply agency has to supply good quality water through tankers, incurring a very high cost, the communities face a lot of hardship in fetching the limited quantities of water supplied through such arrangements. Often households spend their own resources to purchase water from private vendors. Such costs need to be considered while estimating the life cycle cost of rural water supply schemes for informed decision-making.

References

Bassi N, Kabir Y (2016) Sustainability versus local management: comparative performance of rural water supply schemes. In: Kumar MD, Kabir Y, James AJ (eds) Rural water systems for multiple uses and livelihood security. Elsevier, Netherlands, UK and USA, pp 87–115

Isham J, Kähkönen S (2002) Institutional determinants of the impact of community-based water services: evidence from Sri Lanka and India. Econ Dev Cult Change 50(3):667–691

Kumar MD, Kabir Y, James AJ (eds) (2016) Rural water systems for multiple uses and livelihood security. Elsevier, Netherlands, UK and USA

MacAllister DJ, MacDonald AM, Kebede S, Godfrey S, Calow R (2020) Comparative performance of rural water supplies during drought. Nat Commun 11. https://doi.org/10.1038/s41467-020-14839-3

Moriarty P, Smits S, Butterworth J, Franceys R (2013) Trends in rural water supply: towards a service delivery approach. Water Altern 6(3):329–349

Pattanayak SK, Poulos C, Wendland KM, Patil SR, Yang JC, Kwok RK, Corey CG (2007) Informing the water and sanitation sector policy: case study of an impact evaluation study of water supply, sanitation and hygiene interventions in rural Maharashtra, India. Working paper 06_04. Research Triangle Institute International, Research Triangle Park, NC

Prokopy LS (2005) The relationship between participation and project outcomes: evidence from rural water supply projects in India. World Dev 33(11):1801–1819

Reddy VR, Rammohan Rao MS, Venkataswamy M (2010) 'Slippage': the bane of drinking water and sanitation sector (a study of extent and causes in rural Andhra Pradesh). WASHCost India-CESS working paper. Hyderabad

WHO, UNICEF (2006) Meeting the MDG drinking water and sanitation target: the urban and rural challenge of the decade. World Health Organization and UNICEF, Switzerland

Wilbers GJ, Sebesvari Z, Renaud FG (2014) Piped-water supplies in rural areas of the Mekong Delta, Vietnam: water quality and household perceptions. Water 6(8):2175–2194. https://doi.org/10.3390/w6082175

World Bank (2008) Review of effectiveness of rural water supply schemes in India. Report 44789. https://openknowledge.worldbank.org/bitstream/handle/10986/19488/447890ESW0WHIT10Bank0Report10July02.pdf?sequence=1&isAllowed=y. Accessed 15 Aug 2021

Chapter 5
Locating Water for Augmenting Rural Water Supply Schemes

5.1 Introduction

During the past three decades or more, there have been a number of initiatives by the government agencies at the state and central levels to implement small water harvesting and artificial groundwater recharge schemes under various programmes. While some of them were under local water conservation programmes, some of them were done under the Integrated watershed development programme and some under National Rural Employment Guarantee Scheme (NREGS). In a few instances, they were undertaken for strengthening groundwater-based drinking water supply schemes. The successful implementation of all these programmes, be it for water harvesting or artificial recharge of groundwater, required a sufficient amount of run-off water to be made available from the local catchments (where they are proposed to be built).

What is common about all these programmes is that they were all planned in a decentralized way by local agencies such as the District *Panchayats*, District Rural Development Agencies, and Village *Panchayats*, with financial support from either the state government departments or the central government agencies but without any technical expertise that is required for planning and designing water infrastructure. Prior to planning such works, no assessments of total water available in the river basins or even in the local watersheds were undertaken (Kumar et al. 2008). No investigations were carried out for understanding the terrain conditions vis-à-vis geology, geohydrology, topography, and soils of the area that provide critical inputs for selection of sites, appropriate structures, and hydraulic and structural design.

As a result, generally, no scientific data or technical data are available on dependable run-off (yield) of the catchments, the peak discharge that occurs during high-intensity storms that are required for deciding the storage capacity that can be created, and designing the spillways to evaluate floods, respectively. The departments that carry out these works have no hydrologists or water engineers who can do proper planning of the schemes based on resource assessment and physical conditions (topography, drainage characteristics, and geological conditions) of the catchments. They

M. Dinesh Kumar et al., *Drinking Water Security in Rural India*, Water Resources Development and Management, https://doi.org/10.1007/978-981-16-9198-0_5

generally look at these works as civil construction works. The argument often put forth by these departments against following such a technically sound approach is that the structures being planned are too small to cause any negative effect on the stream flows.

What is more worrisome is the fact that the volume of committed flows from these catchments for the ongoing schemes are not taken cognizance of, as the agencies that execute such schemes do not coordinate with the agencies that are responsible for managing the existing systems, the latter being the State Water Resources Departments. Since the schemes are often small and carried out at a small scale by the local self-governing institutions such as the Village *Panchayats*, they are below the radar of the latter. As a result, works carried out in the past 2–3 decades without any scientific and technical inputs had led to several problems. They are (1) adverse impact on existing downstream uses with reduced inflows into reservoirs and tanks (Ray and Bijarnia 2006; Kumar et al. 2008; Glendenning et al. 2012); (2) over-designed structures which harness water to fulfil their storage capacities only once in many years (Kumar et al. 2006, 2008); and, (3) poor siting of the structures (Bassi and Kumar 2010) leading to quick damage during monsoon due to bank erosion, over-topping, piping, etc.

Therefore, whenever new schemes, whether large or small, are planned, the biggest technical challenge is to do a proper assessment of the catchment to understand the yield characteristics (dependable yield, yield variations between years, etc.) along with the current extent of utilization of water from the catchment (or the committed flows). Only such assessment can indicate how much water is available for future harnessing and in what percentage of the years. Here, one should remember that it is not the size of the structure which matters, but the cumulative effect of several structures when built in a catchment on the hydrology and water systems. It is all the more important for regions that are physically water-scarce and where large-scale interventions for water resources development (reservoirs of all sizes) already exist. Such studies will be crucial for planning reservoir-based schemes for drinking water supply in India's villages to ensure sustainable water supply in the future.

In this chapter, we present the methodology for estimation of un-utilized water resources in situations where the river basins have several large-, medium-, and small-sized water resource systems that divert water for various uses (major, medium, and minor irrigation reservoirs). This methodology was then used to estimate the un-utilized surface water resources in major river basins of Maharashtra that can be utilized for augmenting existing or developing new water supply schemes. For this purpose, a systematic assessment of water utilization by major and medium and minor reservoir and diversion-based schemes in each basin was made, and the estimates were compared with the total renewable water resource available in those basins. Further, the current proposals for improving the sustainability of rural water supply were analysed and compared against the analysis of water supply schemes' performance (see Chap. 4) to determine their usefulness in achieving domestic water security.

5.2 Building the Case for Reservoir-Based Water Supply Schemes

It is clear from the analyses presented in Chap. 3 that there are many districts and parts of certain districts in Maharashtra where groundwater-based schemes will not be able to provide dependable water supply for drinking purpose in the villages. The primary data collected from sample households served by two distinct types of schemes (i.e. groundwater and surface water, with one representative scheme from both types), in each of the six divisions of the state also suggest that the reservoir-based schemes perform better (refer to Chap. 4). The results of the primary data analysis presented in Chap. 4 also seem to suggest that when served by reservoir-based schemes the households show a greater tendency to go for piped water supply connection inside the dwelling and within the dwelling premise and also adopt modern toilets, a clear indication of the guarantee for source water quality and dependability.

The areas where groundwater-based schemes are not sustainable will have to be provided with surface water since in the case of surface water schemes the resource largely remains under the control of the state water resources department and it is possible to limit the access of private individuals to the resource in most instances through proper surveillance. The effectiveness of this control becomes greater for reservoirs as compared to rivers and streams. The presence of reservoirs makes it easy to quantify the total amount of water that is available for supply soon after the monsoon (as most of the inflows arrive during the monsoon itself). Therefore, the water managers can plan its utilization, the demand for which starts after the monsoon. It is even possible to physically allocate water from the reservoir in volumetric terms for different uses, and this is normally done for most multi-purpose reservoirs.

In the case of schemes that are dependent on lifting water directly from the river, the reliability is generally poor, as it is difficult to forecast the lean season flows in the river. So, if river lifting schemes are planned, then they should be resorted to only when the river carries controlled releases from reservoirs upstream. Or else, it should be ensured that the lean season flow in the river has the least inter-annual variability; i.e., it does not vary widely between years.

Generally, when drinking water schemes are built around surface water sources, they are planned for several villages and not for one hamlet or village, unlike in the case of groundwater-based schemes such as handpumps, bore wells, and tube wells that are built to serve a group of households within a village or a hamlet. The reason for this is that we achieve economies of scale when a large number of villages are covered. The reason is that the capital cost of the common infrastructure (water lifting device, raw water treatment plant, water mains, etc.) is generally very high for reservoir-based schemes, irrespective of the size of the population to be served, as the villages to be served from the scheme are located at far-off locations. By increasing the number of villages to be served, this cost gets distributed to a larger population, bringing down the cost per unit volume of water supplied or the cost per HH served. The number of villages can go up to a few hundred in the case of regional water supply schemes based on reservoirs. However, the total number of villages considered for planning such regional water supply schemes depends on the total amount of water available in the source. Hence, it is important that when regional water supply schemes are planned for drinking water supply, large sources of water having highly dependable yield are chosen. Such considerations help improve the cost-effectiveness and long-term sustainability of the drinking water scheme.

5.3 River Basins of Maharashtra

Five large river basins have their catchment area, either partially or fully, in Maharashtra. These are Godavari, Krishna, Mahanadi, Narmada, and Tapi. In addition, several West Flowing rivers also have their catchments in the state. The drainage of these basins altogether forms 26 sub-basins in the state. The boundaries of the major river basins of Maharashtra and their sub-basins are shown in Fig. 5.1. With respect to their catchment area in Maharashtra, Godavari, Krishna, Tapi, and the West Flowing rivers are the major river basins in the state. Two other basins, namely Narmada and Mahanadi, have only a small portion of their catchment in Maharashtra.

Several of the sub-basins of the two large river basins, viz. Godavari, and Krishna—in terms of the drainage area of the basin falling inside Maharashtra—face severe water shortage, while very few are water surplus or water abundant. Overall, the demand for water for agriculture in these two basins is high in comparison with the total amount of renewable water resources generated annually, owing to the large amount of arable land and the semiarid climatic conditions. In other words, the total amount of renewable surface water resources available per unit area of arable land is quite low in many of the sub-basins, while water demand for agriculture per unit of cultivable land remains high. Historically, there has been a high degree of development of water resources for use in irrigation and other competitive use sectors such as municipal use and industry in these basins. The amount of surface water available for future exploitation is low (as per the allocation by water dispute tribunals) except in wet years. There are very few rivers originating in the Western Ghats basins which are either water surplus or water abundant. Not only that the surface run-off availability is very high, but the demand for water for irrigation and other competitive uses is quite low in these basins as the proportion of the land area suitable for cultivation is very low.

The water resources department of the government of Maharashtra had classified various river basins and their sub-basins falling inside in the state on the basis of renewable surface water availability per unit of cultivable command area (CCA) into five categories as 'water abundant', 'water surplus', 'normal', 'water deficit', and 'highly water-deficit'. The criterion used for classification are as follows: below 1500 m^3/ha is highly deficit; 1500–3000 m^3/ha is deficit; 3000–8000 m^3/ha is normal; 8000–12,000 m^3/ha is surplus; and above 12,000 m^3/ha is abundant. Accordingly, the different categories in which each sub-basin/river basin falls are given in Fig. 5.2.

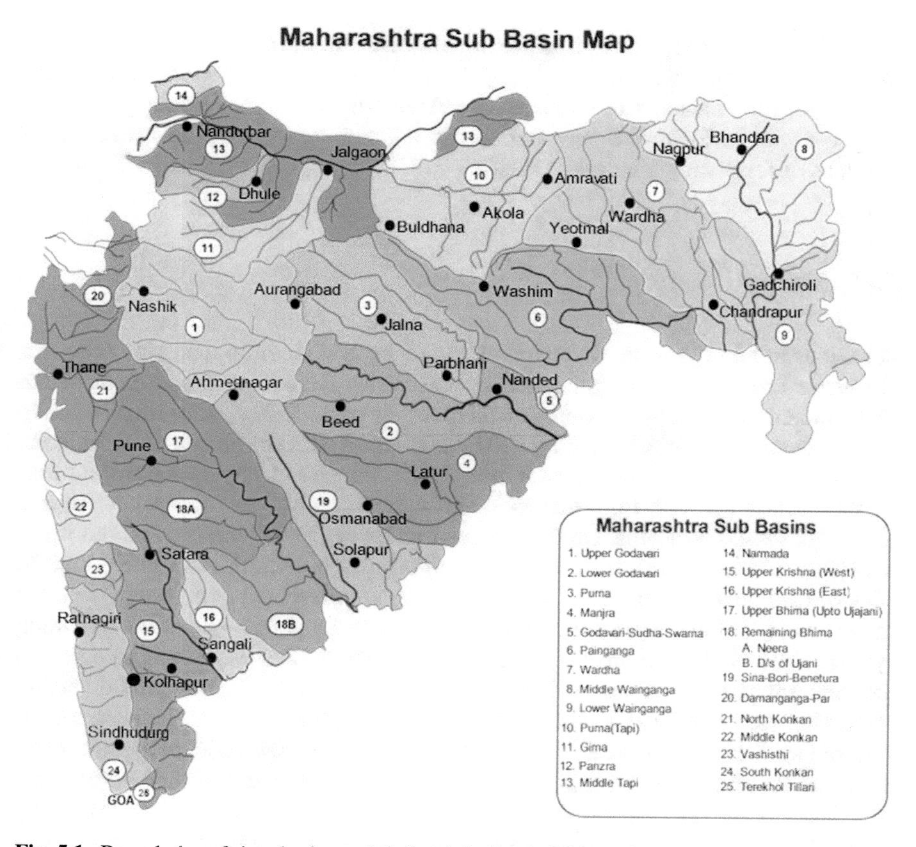

Fig. 5.1 Boundaries of river basins and their sub-basins in Maharashtra

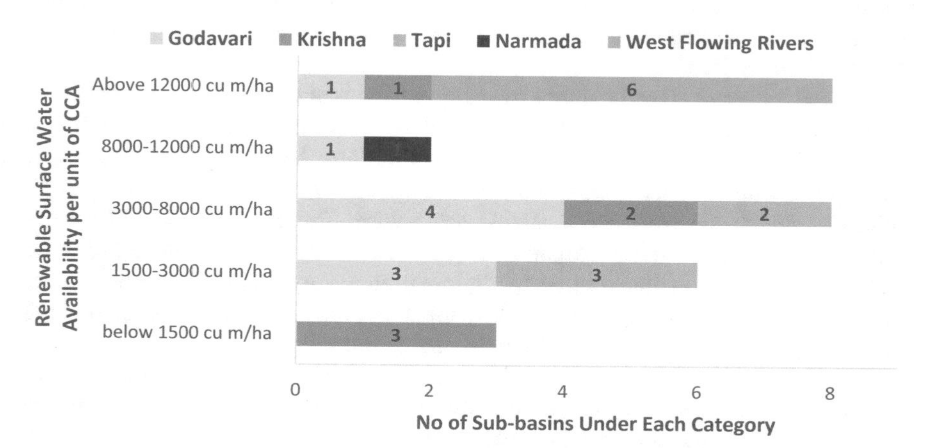

Fig. 5.2 Classification of sub-basins/river basins according to water availability. *Source* Government of Maharashtra (2012)

While the foregoing description provides the agricultural demand scenario against the renewable water resources, it is important to get a realistic picture of the actual water utilization scenario for these basins and the water that remains untapped, which will be available for future exploitation for augmenting the existing water supply schemes or for planning new schemes.

5.4 Methodology

For the estimation of water diversion by the major and medium irrigation schemes, the water accounting data generated by the Water Resources Department of the Government of Maharashtra were collected for the period from 2008–2009 to 2017–2018 (except that of 2010–2011 and 2011–2012) and used for further analysis. Each such report provides data on the reservoirs design storage capacity, actual live storage, water releases for irrigation, domestic and industrial purposes, irrigation projects level water uses, water lost from the reservoirs through evaporation, and the water which remained un-utilized in the reservoirs during a hydrological year (June to May of next year). All the data and information are presented both at the irrigation circle level (administrative unit for the operation and maintenance of the irrigation systems) and at the sub-basin/river basin level. These data were used to estimate the total amount of water effectively diverted through the reservoirs and diversion systems of the 71 major schemes and 255 medium irrigation schemes in the state. For this, the total amount of water diverted for various uses, reservoir evaporation and the water that remained un-utilized in the reservoirs at the end of the hydrological year were considered. The estimates of total effective water diversion were made by taking the average of the figures for the eight years. Since the water accounting data for medium schemes were available for only 255 out of the 297 schemes, the total utilization was arrived at using extrapolation, by multiplying the utilization figures by a fraction, 297/255.

For water diversion by the minor irrigation schemes and the river lift schemes run by the farmers, data from the water accounting report of 2017–18 were utilized. Though there are 3516 minor irrigation schemes in the state as per the water accounting report of 2017–18, water accounting data were available for only 2213 schemes, which too were not available basin-wise. Therefore, the average water utilization by a single minor irrigation scheme (based on water accounting data for 2213 schemes) was estimated first by dividing the water utilization by the number of schemes. This value was multiplied by the total number of such schemes (3519) to estimate the total diversion by the minor irrigation schemes.

In order to estimate the water utilization by river lift schemes, for which water accounting data were not available, the following methodology was used. First, the average water utilization per ha of irrigation for minor schemes was estimated by dividing total water utilization by the gross irrigated area by such schemes. It was assumed that the same delta will be maintained by river lift schemes. Thereafter, the unit delta was multiplied by the gross irrigated area under the river lifting schemes to estimate the total water diversion from such schemes.

Since basin-wise irrigated area or number of schemes were not available for minor schemes and river lift schemes, the total water utilization by these two different types of schemes was apportioned among the basins on the basis of the irrigated area by the major and medium schemes in those basins. It was assumed that the basin-wise area irrigated by the minor and lift irrigation schemes is also in the same proportion as that of the major and medium schemes.

The effective water diversion from the major, medium, minor, and river lift irrigation schemes was used to estimate the total amount of water available in these basins that remain untapped and which will be available for future exploitation. For this, the water already diverted effectively through various schemes was subtracted from the total amount of renewable water resources of the basins that are allocated to the state of Maharashtra through various tribunal awards. Only four river basins, namely Godavari, Krishna, Tapi, and west flowing rivers, were considered for the analysis. The two other river basins, i.e. Narmada and Mahanadi, have only a small portion of their catchment area in Maharashtra and hence are not considered for the analysis.

5.5 Surface Water Availability in Selected River Basins

5.5.1 Water Utilization by Minor Irrigation Schemes and River Lift Schemes

As per the estimates based on the water accounting report of 2017–18, the total water utilization (water diversion and losses) by the 2213 minor irrigation schemes was 3530 million cubic metres (MCM). The average water utilization by a single minor irrigation scheme comes out to be 1.595 MCM (3530 divided by 2213). Thus, the estimated water utilization by the total 3519 minor irrigation schemes was 5614 MCM (3519 multiplied by 1.595).

Further, the gross irrigated area by the minor irrigation schemes was 0.3464 million hectares (m. ha) in 2017–18. Thus, the average water utilization per ha of irrigation for minor schemes was estimated to be 0.0162 MCM (5614 divided by 0.3464×10^6). The gross irrigated area by all river lifting schemes was 0.1483 million hectares (m. ha) in 2017–18. Using the delta (irrigation depth) of the minor irrigation schemes and the gross irrigated by the river lift schemes, the water utilization by such schemes was estimated to be 2403 MCM ($0.0164 \times 0.1483 \times 10^6$).

Thus, overall, 8017 MCM of water is utilized by the minor irrigation schemes and the river lift irrigation schemes operated by the farmers. The total water utilization by these two different types of schemes was apportioned among the basins on the basis of the irrigated area by the major and medium schemes in those basins using an assumption that the area irrigated by these schemes in different basins is also in

Table 5.1 Estimates of river basin-wise water utilization by minor and river lift schemes, Maharashtra

Name of the river basin	Average irrigated area by medium and major irrigation schemes (ha)	% to the total irrigated area	Total estimated water utilization by minor schemes and river lifting schemes	Estimated water utilization by the minor and river lifting schemes in different basins (MCM) based on apportioning
Godavari	515,412.95	29		2336.47
Tapi	95,186.14	5		431.50
Krishna	1,146,493.71	65		5197.30
West flowing rivers	11,411.69	1		51.73
Overall	1,768,504.49	100	8017.00	8017.00

Source Authors' estimates based on 2017–18 data

the same proportion as that of the latter. The outputs with regard to basin-wise total water utilization by all minor and RL schemes are presented in Table 5.1.

5.5.2 *Water Utilization by Medium and Major Irrigation Schemes*

Over the period from 2008–2009 to 2017–2018, the estimated average annual effective water diversion including the evaporation losses from the rivers through the 297 medium and 71 major irrigation projects was about 24,719 MCM. River basin-wise water utilized by such schemes is provided in Fig. 5.3. It should be noted that the effective diversion changed from year to year, depending on the inflows available to the schemes and demand for water, especially for irrigation.

5.5.3 *Un-utilized Water Resource in the Basins*

The final outputs with regard to water utilization by minor, medium, major, and river lift irrigation schemes, and the total water available for further utilization in different river basins is given in Table 5.2. Comparing the results with the information presented in Fig. 5.2 shows that once developed the un-utilized water available in those basins will help mitigate the scarcity situation prevailing in those basins to some extent. On the other hand, in the river basins of west flowing rivers, there is plenty of un-utilized water resources which can be tapped for use in other basins, which have a large amount of arable land lying uncultivated due to water shortage.

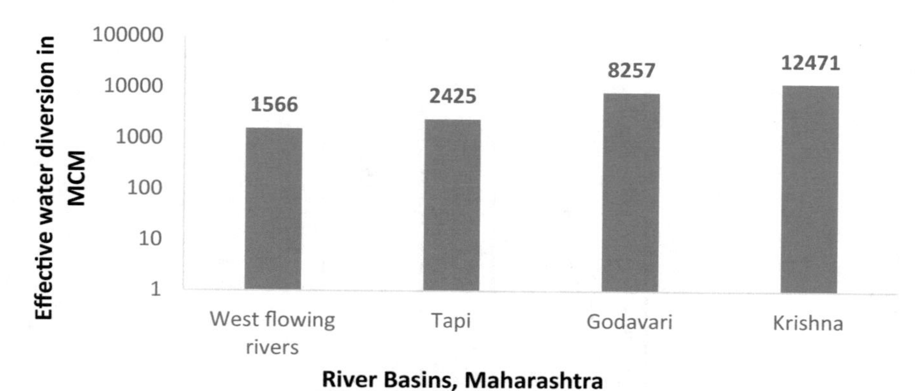

Fig. 5.3 Estimates of river basin-wise water utilization by medium and major irrigation schemes, Maharashtra. *Source* Authors' estimates using water accounting reports from 2008–09 to 2017–18

The largest amount of water is in the west flowing rivers, amounting to 62,600 MCM per annum. The second largest amount is in the Godavari basin, 18,430 MCM.

In the case of the Krishna River basin, the figure is negative. Practically, this is not possible in the sense that the actual water utilization cannot be more than renewable surface water resources. This is because we have considered the 75% dependable yield (i.e., the minimum flow that occurs in 75% of the years) of the basin and the state's allocation corresponding to that (16,220 MCM) and not the actual yield, and therefore, the actual allocation that the state is entitled to. The actual yield will be more than the dependable yield for 75% of the years. The utilization data suggests that during the years considered for the analysis, the actual flow in the Maharashtra part of the Krishna River basin was more than 17,668 MCM. Ideally, it should also have been more than the dependable yield of 22,299 MCM which had enabled Maharashtra to use more than its allocation of 16,220 MCM. Using the data sets presented here, we cannot estimate what the actual flow was. For that, we need to analyse the data at the last drainage point in the Maharashtra part of the Krishna River before it enters Telangana. But that is out of the scope of the current work.

5.6 Current Proposals for Improving Rural Water Supply in Maharashtra

In the past, the Government of Maharashtra (GoM) has implemented small-scale water harvesting and groundwater recharge programmes to augment local water availability for boosting the domestic water supply in rural areas. However, augmenting groundwater through local water harvesting and recharge schemes will only help augment water supply for irrigation, that too in a very limited way. The reasons are (1) in the low to medium rainfall regions (like Marathwada and Vidarbha), the

Table 5.2 Estimates of water utilization and un-utilized flows in different river basins of Maharashtra

Basin/sub-basin	Catchment area (km^2)	Culturable area (m ha)	Surface water availability (MCM)		Water utilization (MCM)		Un-utilized water (MCM)
			Natural surface water availability (75% dependability)	*Water allocation as per tribunal award*	*Major and medium irrigation projects*	*Minor and river lift schemes (MCM)*	
Godavari	151,122	11	38,607	29,023	8257	2336	18,430
Tapi	52,058	3	7027	5996	2425	432	3140
Krishna	69,425	6	29,299	16,820	12,471	5197	– 848
West flowing rivers	31,507	1	64,218	64,218	1566	52	62,600
Total	304,112	21	139,151	116,057	24,719	8017	

Source Authors' own estimates

amount of water that can be tapped for recharging groundwater from the catchments is very meagre (due to very limited run-off) in comparison with the supply-deficit in irrigation, and (2) in the high rainfall hilly and mountainous regions (western Maharashtra), the aquifer recharge potential is very low due to unfavourable geology and geomorphology (refer to Chap. 2). In any case, such efforts will not help augment the drinking water schemes when it is needed the most (i.e. during summer), due to the precarious demand pattern.

Moreover, as studies have shown, water harvesting and groundwater recharge that have been undertaken during the past few years only resulted in reduced inflows into many surface reservoirs downstream (Ray and Bijarnia 2006; Kumar et al. 2008). Therefore, from a river basin perspective, there is no overall gain from water harvesting and recharge schemes in the water-scarce regions (Kumar et al. 2008). When the upstream communities gain some extra water, the downstream communities lose. Notably, many of the villages that have witnessed decentralized water harvesting are dependent on tankers during the lean season.

Further, the groundwater regulation enforced by the state of Maharashtra is largely ineffective in checking groundwater pumping for irrigation and improving water availability for drinking and domestic uses in rural areas due to competing priorities and many other vested interests in the villages. The first of such legislation, the Maharashtra Groundwater (Regulation for Drinking Water Purposes) Act, 1993 had several constraints and drawbacks. These include provisions of the Act were only enforceable either in watersheds declared as over exploited or in a specific locale notified as scarcity affected in a particular year; the Act was silent over any over exploitation of wells located beyond 500 m from a public water source; the Act did not make it compulsory to register the irrigation wells; and only one-time punitive action was provided for the village for particular violation in the respective year precluding any long-term measures (Phansalkar and Kher 2003).

Subsequently, the Maharashtra Groundwater (Development and Management) Act, 2009 was enacted to address such concerns. The 2009 Act has several new features, such as the state was given more power to manage groundwater by making Maharashtra Water Resources Regulatory Authority (MWRRA) as state Groundwater Authority; prohibition on drilling and extraction from deep bore wells; no permission for the construction of new wells within the notified areas; and focus on source as well as resource management, undertaking groundwater accounting, crop planning, preparing ground water use plan, decentralization and community participation, maintenance of strict water quality measures, and mapping of aquifers. However, the 2009 Act also failed to empower the village *Panchayats* to effectively execute the regulatory norm framed by the state.

Of late, to address the water scarcity concerns, the Maharashtra government has initiated programmes to transfer water from the comparative water-rich river basins (such as Godavari) to water-scarce basins (such as Krishna). Originally, the inter-basin transfer of water transfer is the initiative of the then Ministry of Irrigation (now Ministry of Jal Shakti) in 1980. For this purpose, the Government of India (GoI) created National Water Development Agency (NWDA) in 1982 to identify the river basins that are water surplus and water-deficits in order to explore the possibility of storage, links and water transfers. Based on the detailed hydrological analysis, NWDA identified 30 river links for transferring water from the comparative water surplus to water-deficit river basins in the country and also developed feasibility reports on some of them.

Some of the benefits of linking river basins are flow control, low cost and best water for irrigation, availability of drinking water, generation of hydroelectric power, and employment generation. However, the inter-basin transfer of water is hindered for one or other reasons. The main concerns are related to adverse impacts on the environment and ecology and the social cost of human and livestock displacement (George et al. 2014). Recently, the Supreme Court of India had given directives to fasten the projects on inter-basin transfer of water and also to undertake a proper assessment of the environmental and social damages that should be adequately compensated.

For Maharashtra, two inter-basin water transfer projects that are Par-Tapi-Narmada and Damanganga-Pinjal link project are considered as important. In these two projects, surplus water from the west flowing river basins is proposed to be transferred to the Tapi basin in Maharashtra and Gujarat (using the Ukai Dam reservoir) and then to the Narmada basin in Gujarat. The main purpose of such inter-state projects is to provide water for irrigation and hydropower generation. In addition, the state government has proposed eight other inter-basin and two intra-basin links. The details of the proposed links are provided in Table 5.3.

Subsequently, nine more links were identified and proposed under the Girna-Basin, Anjani, Tittur, and Aadnadi river links project in Maharashtra. A total of 127 MCM of water will be diverted for drinking and irrigation purposes at the cost of INR 112.15 million (Patel et al. 2018). This project will benefit drought prone areas in the Jalgaon district of Maharashtra, and its technical features are presented in Table 5.4.

Thus, the state government is increasingly moving towards solving drinking water crisis in rural areas by inter-basin water transfer. A hydrological model set up to assess probable change in water availability once the water will be transferred from

Table 5.3 The quantum of water diversion proposed in various inter-basin transfer links, Maharashtra

Inter-basin links	Quantum of water diversion (in MCM)
Upper Ghat-Godavari Valley (Damanganga (Ekdare)-Godavari Valley)	143
Upper Vaitarna-Godavari Valley	136
North Konkan-Godavari Valley	269
Koyna-Mumbai city	1912
Middle Konkan-Bhima Valley	425
Savitri-Bhima	708
Nar-Par-Girna Valley	534
River linking projects of Tapi basin and Jalgaon district	74

Source NWDA, GoI

Table 5.4 Details of proposed Girna-Basin, Anjani, Tittur and Aadnadi links project, Maharashtra

Proposed link	Technical features
Interlinking Girna and Bori river	Girna major irrigation project is the donor reservoir that will supply its surplus flows (20.8 MCM) through Panzar left bank canal to Bori River. The water will travel for about 68 km. The storage capacity of the Bori Dam is 1400 million cubic feet (MCFT). The water from the Bori Dam will be used to supply water to one town and 73 villages in the Jalgaon district
Filling of Matera and Bhokarbari project	Runover water from the Girna reservoir will be diverted to Jamda left canal connected to Parola branch canal and finally reaching to Bhokarbari Dam. The total volume of water that will be diverted is 19.5 MCM and it will travel a distance of 132 km. This water will be supplied for drinking purposes
Interlinking Girna and Anjani rivers	Excess water (about 18.1 MCM) from the Girna reservoir will be released into the Anjani river (at a distance of 127 km) which will be then supplied to one town and several villages en-route for drinking and other domestic purposes
Interlinking Girna and Tittur rivers	Excess water (about 14.3 MCM) from the Girna Dam will be transferred to Tittur River (at a distance of 84 km) which will be supplied for drinking and irrigation purposes
Filling Pimpri dam	Excess water (about 13.1 MCM) released from the Girna Dam will be chanelled through the Girna lower canal to Pimpri Dam (at a distance of 1186 km) to provide water for drinking and irrigation purposes
Panzan left bank canal to Girna River	The flood water from the Panzan left canal (about 19.1 MCM) will be chanelled to Girna River
Filling Kadgaon Dam	The excess water (about 5.2 MCM) from Girna reservoir will be diverted to Jamda right canal and transferred to Kadgaon Dam traversing about 76 km to provide water for drinking water and irrigation
Filling of Patna and Kodgaon Dam	The overflow water (about 9.4 MCM) from the Aad River will be transferred to the Kodgaon reservoir. After supplying water for drinking and irrigation purposes, the leftover water will be returned to Aad River
Linking Aad, Dongri, and Tittur rivers	The overflowing water from the Aad River will be transferred to the Chitegaon Dam covering a total distance of 15 km. This water will be used for drinking and irrigation

the Dhom reservoir to Ner reservoir in Maharashtra found that such transfer will improve the water balance in the receiving area and make more water available to the nearby community, thereby improving their livelihoods (Munde et al. 2017).

Further, a techno-economic feasibility analysis of supplying surface water to a cluster of 17 villages in the Thane district of Maharashtra that were previously dependent on tanker water supply found that a multi-village scheme based on surface water reservoir located at the higher elevation can reduce both capital cost of piping and the energy cost of pumping water by allowing gravity flow of water (Hooda et al. 2013). The per capita cost of supplying water to the cluster of 17 villages was estimated to be INR 2888 which is lower than the government norm of INR 3250. Further, the annual operation and maintenance (O&M) cost including energy cost was estimated at INR 6.35 per cu. m. of water, which is much below the government norm of INR 16. Similar results were obtained from the other such multi-village surface water-based schemes in the state (WSP 2004).

5.7 Conclusion

The analysis presented in this chapter confirms the availability of surplus water in some of the river basins of Maharashtra, including the Godavari and many of the west flowing rivers originating from the Western Ghats. Also, there are many reservoirs in the state, including those in water-scarce basins which have un-utilized stock at the end of the summer season. Yet there are thousands of villages that face drinking water shortages during summer months and are dependent on tanker water supply. The scope for developing or utilizing this water for augmenting water supply schemes in the region needs to be studied. Also, in some cases, there might be scope for reallocating some portion of the water from many of the existing reservoirs for drinking water supplies. A broad strategy considering such aspects for achieving sustainable rural domestic water supply in Maharashtra will be discussed in the next chapter.

Thus, for Maharashtra to make progress towards providing ‘tap water’ to every household in the next few years, the only viable solution that appears is to go for a hybrid model of surface and ground water, except in the Konkan region which has high rainfall and alluvial formations where wells as a source of drinking water are still viable. Apparently, the region does not experience a shortage of water for irrigation and domestic uses. The districts of the Konkan region also have relatively higher human development indices, with better economic conditions, health, and literacy. This will ensure the sustainability of the ‘source water’, as it is easy for the water supply departments of the states to allocate a portion of the water from surface water sources for domestic use, which is not possible in the case of groundwater due to the issue of ownership. Further, the biggest advantage of using surface water for drinking water supply is that the raw water does not require expensive treatment, as the surface water is free from chemical contaminants.

References

Bassi N, Kumar MD (2010) NREGA and rural water management in India: improving the welfare effects. Occasional paper no. 3. Institute for Resource Analysis and Policy, Hyderabad

George CM, Korgaonlear DP, Geetha K (2014) Interlinking of river basins: a review. Int J Civ Struct Environ Infrastruct Eng Res Dev 4(2)

Glendenning CJ, Van Ogtrop FF, Mishra AK, Vervoort RW (2012) Balancing watershed and local scale impacts of rain water harvesting in India—a review. Agric Water Manag 107:1–13

Government of Maharashtra (2012) Report on water auditing of irrigation system in Maharashtra state 2010–11. Water Resources Department, Government of Maharashtra, India

Hooda N, Desai R, Damani OP (2013) Design and optimization of piped water network for tanker fed villages in Mokhada Taluka. Indian Institute of Technology, Mumbai

Kumar MD, Ghosh S, Patel A, Singh OP, Ravindranath R (2006) Rainwater harvesting in India: some critical issues for basin planning and research. Land Use Water Resour Res 6:1–17

Kumar MD, Patel A, Ravindranath R, Singh OP (2008) Chasing a mirage: water harvesting and artificial recharge in naturally water-scarce regions. Econ Pol Wkly 43(35):61–71

Munde S, Vaidya D, Nikam S, Sathe N (2017) Linking of Dhom reservoir to Ner reservoir: understanding the alignment of the open and close conduit, Maharashtra. Int J Res Publ Eng Technol 3(10)

Patel NA, Suryavanshi DS, Jaidev BR (2018) Strategic analysis of interlinking of rivers in drought prone tehsils of Jalgaon district—Maharashtra. Asian J Sci Technol 9(1)

Phansalkar SJ, Kher V (2003) Decade of Maharashtra groundwater legislation: analysis of the implementation process in Vidarbha. International Water Management Institute, Colombo

Ray S, Bijarnia M (2006) Upstream vs downstream: groundwater management and rainwater harvesting. Econ Pol Wkly 41(23):2375–2383

WSP (2004) Focus on Maharashtra: alternate management approaches for village water systems. World Bank, Delhi

Chapter 6
Strategies for Improving Rural Domestic Water Supply in Maharashtra

6.1 Current Discourse for Water Supply Augmentation

India has finally begun to witness a paradigm shift in the rural water supply. The notion among policymakers of what it takes to supply clean water to rural areas on a sustainable basis year-round seems to be changing. As discussed in Chap. 3, for several years, policymakers at the state as well as central level were obsessed with local and decentralized options for rural water supply. The constitutional amendment that emphasized decentralization for the governance of public services, to a great extent, influenced this thinking at the policy level. In spite of growing evidence of the failure of groundwater-based schemes, inability to find adequate amounts of resources to invest in large, multi-village schemes, the requirement of comprehensive regional water planning that demanded time and effort, and the absence of any working model to emulate prevented the agencies from looking at long-term and sustainable solutions for rural water supply.

However, things have changed since the early 2000s. The positive experience of the water supply department of the Government of Gujarat in implementing a large-scale regional water supply scheme costing several billions of dollars and catering to thousands of villages extending over a very large geographical area that involved the bulk transfer of water from Narmada and Mahi canals (Biswas-Tortajada 2014; Jagadeesan and Kumar 2015) had changed the overall outlook of several state governments on regional water supply schemes that involve heavy operation and maintenance. There is a growing realization among many Indian states, which face perennial problems of drinking water shortage, that if the village water supply has to be made sustainable and if the people have to get freedom from the miseries of depending on tankers during summer, the techno-institutional model for achieving that can often be very sophisticated, requiring long-term investments, massive infrastructure, and large resource commitments, especially in regions that are naturally water-scarce. The high 'slippage' witnessed by many states that depend on bore wells for village water supply (Chaudhary et al. 2020; Reddy et al. 2010) had made a compelling case

M. Dinesh Kumar et al., *Drinking Water Security in Rural India*, Water Resources Development and Management, https://doi.org/10.1007/978-981-16-9198-0_6

for looking at the life cycle cost of schemes that otherwise are not capital-intensive (Reddy et al. 2012).

The recent negative experience of Maharashtra in dealing with the drought situation in Marathwada wherein the state government had to transport water through rail wagons from the distant reservoirs located in western Maharashtra (Kumar 2018) had strengthened the arguments in favour of bulk water transfer and elaborate distribution networks for water supply in rural and urban areas. The state of Telangana in southern India had already gone for investing in a very large, state-wide scheme for rural water supply called 'Mission Bhagiratha' that eventually taps only a single source, i.e. Godavari River, as the water supply source. Single village schemes are going out of fashion in rural water supply, especially in the hard rock regions where finding water for domestic water supply during summer is increasingly becoming a complex challenge and areas where groundwater is contaminated with excessive levels of salinity and fluoride (Kumar 2017; Reddy 2018).

Of late, the Maharashtra government has begun contemplating the development of regional piped water supply schemes based on surface water, in an effort to find a lasting solution to the perennial problems of the Marathwada region, that involves connecting large reservoirs within the region for bulk water transfer, and large-scale import of water from water-abundant regions of the state to the water-scarce regions. However, the analysis did not take cognizance of the performance of the existing groundwater-based schemes in different regions of Maharashtra to see where else augmenting/strengthening such schemes would be required for improving their sustainability.

In this chapter, we will systematically examine with the help of regional typologies developed on the basis of certain physical (hydrological, geohydrological, climatic, geomorphological features) and socio-economic (gross and net cropped area, the total area under gravity irrigation, the extent of dependence on groundwater-based schemes for rural water supply) attributes of each district, which regions of the state will be able to provide sustainable rural water supply based on local groundwater-based sources, which regions would require augmentation of the sources from surface water from within the region or outside, and which regions would need to depend entirely on bulk water imports for ensuring drinking water security. For this, it uses the known 'effects' these attributes have on the performance of existing rural water supply schemes. We also estimate the tentative cost of implementing the water supply augmentation schemes and examine the cost-effectiveness of the interventions against tanker water supply.

6.2 Methodology

As a first step, districts in the entire state of Maharashtra were categorized into different typologies on the basis of physical and socio-economic factors. These factors are those which influence effective groundwater availability and demand for groundwater for irrigation, and therefore, the success and failure of water supply schemes have been discussed in detail in Chaps. 2 and 3. They are: recharge per unit area; aquifer storage space; extent of coverage of gravity-based irrigation schemes; irrigation demand per unit area; and extent of dependence on ground-water based schemes for rural water supply. The ten districts that score lowest on each of the four key positive attributes that are critical for successful groundwater-based schemes, viz. sufficient recharge rate per unit area; adequate aquifer storage space; limited irrigation water demand; and increasing the extent of surface irrigation were earlier identified and given in Table 3.4 (of Chap. 3).

Considering the ranking of the districts vis-à-vis each of these attributes, all the districts in Maharashtra were grouped into four clusters: (1) where groundwater-based schemes can sustain for 12 months; (2) where groundwater-based schemes need to be supported by surface water during the summer season; (3) where surface water schemes based on nearby sources will succeed; and (4) where bulk water transfer will be required. Obviously, if a district scores the lowest on all the four attributes mentioned above, it would fall in the fourth category in the sense that it would require imported water for ensuring the sustainability of rural drinking water supply schemes. Conversely, if a district scores the highest on all the four attributes, it would fall in category 1 where groundwater-based schemes will be able to provide year-round water supply. Accordingly, a strategy was proposed to ensure the sustainability of the water supply in all four clusters.

In the next step, the technical feasibility and financial viability of the proposed strategy for improving rural water supply schemes' sustainability were analysed. To ascertain the engineering feasibility of transferring water from the major reservoirs to areas where groundwater-based schemes are either not feasible or need to be augmented in certain parts of the years, a geospatial analysis on the location of major reservoirs in Maharashtra was undertaken. Further, the gross and effective water storage of the reservoirs was analysed to determine which one can be used for the local transfer of surface water and the ones that can be used for the bulk water transfer.

The estimate on the total amount of unutilized water resources in each basin/river system that can be utilized for water transfer was prepared by comparing the total dependable yield of the basins with the total estimated water utilization from major, medium, and minor schemes that are based on reservoirs, diversion systems, and river lifting. The detailed methodology for this is provided in Chap. 5.

For analysing the financial implications of augmenting the water supply schemes from surface sources in different typologies, quick estimates were made on the cost of diverting a unit volume of water from the major and medium projects. For the intra-basin water transfer schemes, first, the total cost (capital and O&M) incurred on all the major and medium projects in Maharashtra up to 2006–07 was used. This cost was annualized (using Eq. 6.1) considering a discount rate of 8% and two different life expectancies for the projects, i.e., 25 years and 50 years.

$$A_{\text{Cost}} = \frac{\text{NPV} \times R}{1 - (1 + R)^{-n}} \tag{6.1}$$

where A_{Cost} is the annualized cost, NPV is the net present value of the asset capital cost and O&M cost, R is the discount rate expressed in fraction, and n is the life of the asset in years. Net present value was estimated using Eq. 6.2.

$$\mathrm{NPV} = \sum_{t=1}^{n} \frac{C_t}{(1+R)^t} \tag{6.2}$$

where C_t is the net cash outflows during a single period t, and t is the total number of time periods.

Then, the annualized cost per unit of area irrigated by the projects in 2006–07 was estimated. Thereafter, using the total volume of water diverted from the projects and the area irrigated, water diverted per ha of the irrigated area was estimated. Subsequently, the annualized cost per cubic metre of water diverted was estimated by taking a ratio of annualized cost per unit irrigated area and water diverted per unit of irrigated area. Finally, this figure was adjusted to the 2016–17 prices using the consumer price index.

For the inter-basin water transfer schemes, capital and operation costs of projects involving a similar scale of water transfer were used to estimate the unit cost of water transfer. The capital cost was annualized (using Eq. 6.1) considering a discount rate of 8% and project life of 50 years. For the groundwater-based schemes, the existing cost of supplying a unit quantity of water from such schemes in the state was considered.

For estimating the total scale of investment required for making rural water supply schemes sustainable, the amount of water to be supplied under different scheme categories was estimated. For this, the projected rural population figures for 2021 (which used decadal growth rates for 2001–2011 and the base population of 2011, obtained from Census 2011) were considered. The per capita water supply considered was 100 L per day. Based on these variables, the total amount of water to be supplied to different clusters in a year was estimated. However, in the case where only water augmentation is required, only the water requirement for three months of the year was considered, as it was assumed that the villages in this cluster would be able to use water from the local groundwater-based schemes for the rest of the year. Hence, the total water demand and the investment requirements could be reduced.

In order to estimate the equivalent one-time capital investment, the net present value of annuity was used. The coefficient is 12.23 for systems with a 50-year life, at a discount rate of 8%. The (annualized) unit cost figures of diverting a unit volume of water are multiplied by this coefficient to arrive at the one-time unit cost. The same is multiplied by the volume of water to be supplied annually to arrive at the total investment required, which also includes the annual operation and maintenance expenses (throughout the life of the scheme adjusted to the zeroth year).

6.3 Technical–Institutional Interventions for Improving the Sustainability of Water Supply Schemes

6.3.1 The Proposal

The multi-pronged strategy for improving the sustainability of rural water supply in Maharashtra is as follows: (1) continuing with groundwater-based schemes in districts and areas where they are sustainable (referred to as Group 1); (2) augmenting groundwater-based schemes with surface water supply from sources from inside or outside the districts wherever possible (conjunctive use schemes) (referred to as Group 2); (3) building schemes based on surface reservoirs in districts/areas where groundwater-based schemes are not successful if sufficient surface water is available locally (referred to as Group 3), and (4) going for bulk water transfer in districts/areas where groundwater-based schemes are not feasible and surface water is extremely limited in the basins in which they fall (referred to as Group 4). The overall approach and strategy are shown in Fig. 6.1.

Group 1 includes Ratnagiri, Raigad, and Sindhudurg districts in Konkan; and areas in the Amravati division along the Tapi river valley. Group 2 includes the districts of Bhandara, Chandrapur, Gadchiroli, Gondia, Nagpur, and Wardha in the Vidarbha region. Group 3 has Kolhapur, Pune, Sangli, Satara, Solapur, and Ahmednagar districts in western Maharashtra. Group 4 mainly comprises districts in the Marathwada region, i.e. Aurangabad, Beed, Jalna, Osmanabad, Nanded, Latur, Parbhani, and Hingoli.

It should be noted that in the case of conjunctive use schemes (Group 2), exogenous water will have to be supplied from reservoirs to the villages only for three months of the year, as it is expected that the groundwater-based water supply schemes will be able to provide water for the rest of the year. However, running such schemes that transfer water for a few months of the year will not be cost-effective, as the capacity of the pipelines that carry the water will be large and the system will remain idle for the rest of nine months of the year. Therefore, water must be transferred throughout the year to existing local storage reservoirs in these Group 2 cluster districts.

Further, the viability of the whole approach and strategy will depend on the following: (1) the geographical location of large surface reservoirs and the engineering feasibility of transferring water from these reservoirs to areas where groundwater-based schemes are either not feasible or need to be augmented in certain parts of the years; (2) the amount of un-utilized surface water resources in different river basins and the engineering feasibility of transferring water in bulk from these basins to areas inside the basin or outside the basin to either strengthen groundwater-based schemes or to build new surface water based schemes; and (3) the cost-effectiveness of the interventions, i.e. how much it costs to develop/appropriate and supply (distribute) a unit volume of water. It is quite obvious that the viability depends on the topography, the location of the reservoirs having surplus water, and their full reservoir levels (FRLs) and distance from places of demand, and how far the basins, where un-utilized surface water is available, are away from the locations of

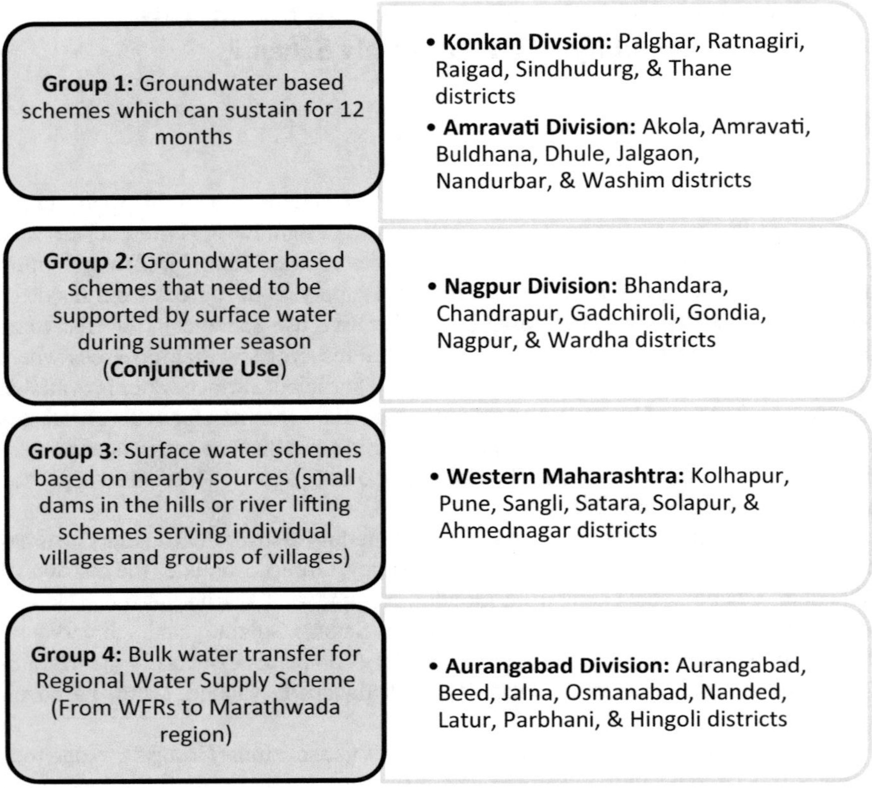

Group 1: Groundwater based schemes which can sustain for 12 months	• **Konkan Divsion:** Palghar, Ratnagiri, Raigad, Sindhudurg, & Thane districts • **Amravati Division:** Akola, Amravati, Buldhana, Dhule, Jalgaon, Nandurbar, & Washim districts
Group 2: Groundwater based schemes that need to be supported by surface water during summer season (**Conjunctive Use**)	• **Nagpur Division:** Bhandara, Chandrapur, Gadchiroli, Gondia, Nagpur, & Wardha districts
Group 3: Surface water schemes based on nearby sources (small dams in the hills or river lifting schemes serving individual villages and groups of villages)	• **Western Maharashtra:** Kolhapur, Pune, Sangli, Satara, Solapur, & Ahmednagar districts
Group 4: Bulk water transfer for Regional Water Supply Scheme (From WFRs to Marathwada region)	• **Aurangabad Division:** Aurangabad, Beed, Jalna, Osmanabad, Nanded, Latur, Parbhani, & Hingoli districts

Fig. 6.1 Proposed approach and strategy for improving rural water supply scheme sustainability in Maharashtra

water demand. The topography, location, and reservoir elevation will determine the need for tunnelling, the distance of water transfer, and the amount of lifting required, all of which have cost implications. The subsequent sub-sections illustrate how the proposed strategy can be implemented.

6.3.2 Engineering Feasibility of Bulk Water Transfer

Geospatial analysis was carried out to determine the location of major reservoirs in Maharashtra, and the results are presented in Fig. 6.2. The purpose of the analysis is to ascertain the engineering feasibility of transferring water from these major reservoirs to areas where groundwater-based schemes are either not feasible or need to be augmented in certain parts of the years. It is clear from the analysis that a large number of reservoirs with capacity in the range of 100–500 MCM and 20–100 MCM

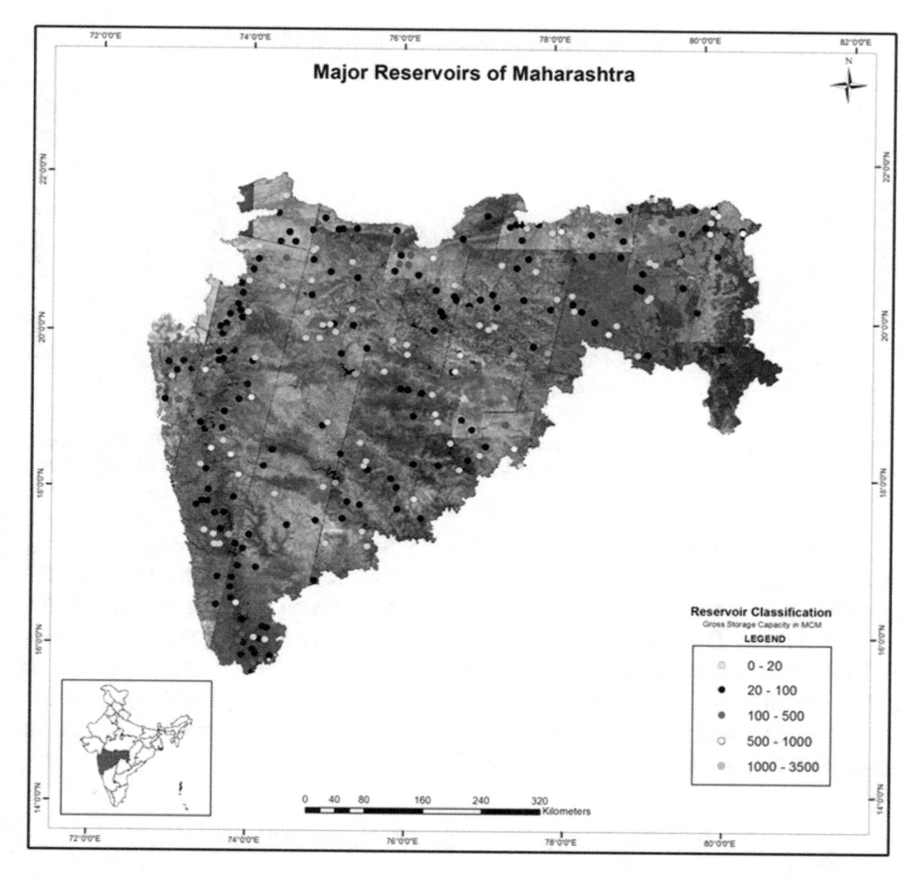

Fig. 6.2 Spatial distribution of major reservoirs in Maharashtra. *Source* IRAP

are on the east-flowing rivers of the Western Ghats in western Maharashtra. Several reservoirs (of varying capacities) are spread across the Marathwada region—a few in the sub-basins of Krishna and many in the sub-basins of Godavari—many (of varying capacities) are located in Vidarbha in the sub-basins of Godavari and some are located in the Amravati region in Tapi basin.

In the next step, the effective water storage capacity of the reservoirs was analysed. In Maharashtra, there are about 2,368 reservoirs for which data on the gross storage are available. The overall gross storage capacity of these reservoirs is 64,130 MCM. Almost 88% of the reservoirs have a gross storage capacity of less than 20 MCM each. Only 3% of the total reservoirs (having storage capacity above 200 MCM) account for 68% of the gross water storage capacity. Figure 6.3 shows the distribution of surface reservoirs in different storage categories and the total gross storage available under each category.

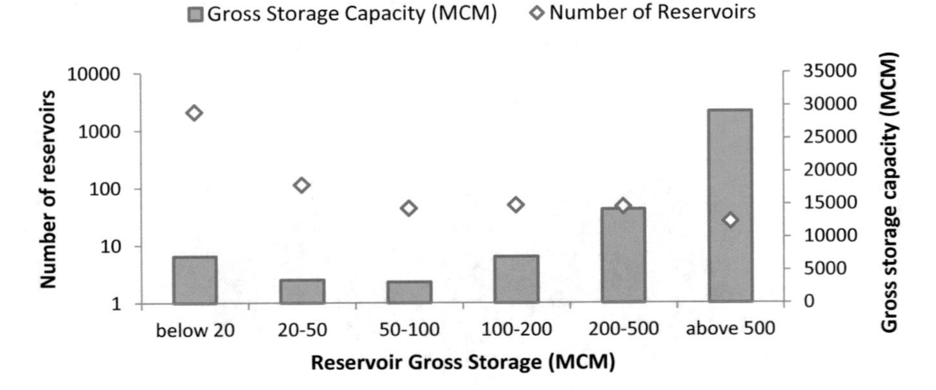

Fig. 6.3 Distribution of reservoirs as per storage capacity and total gross storage. *Source* Authors' analysis using the data from the various water accounting reports of the Government of Maharashtra

In terms of effective storage capacity (i.e. the difference between gross storage and dead storage), data were available for 2340 reservoirs. While 90% of the reservoirs have effective storage availability of less than 20 MCM each, only 3% have above 200 MCM capacity. The latter accounts for 62% of the effective water storage capacity, and the bulk water transfer for regional water supply schemes can be based on them. The smaller capacity reservoirs can be explored for strengthening the existing groundwater-based schemes. The distribution of the number of reservoirs and the gross storage capacity is presented in Fig. 6.4.

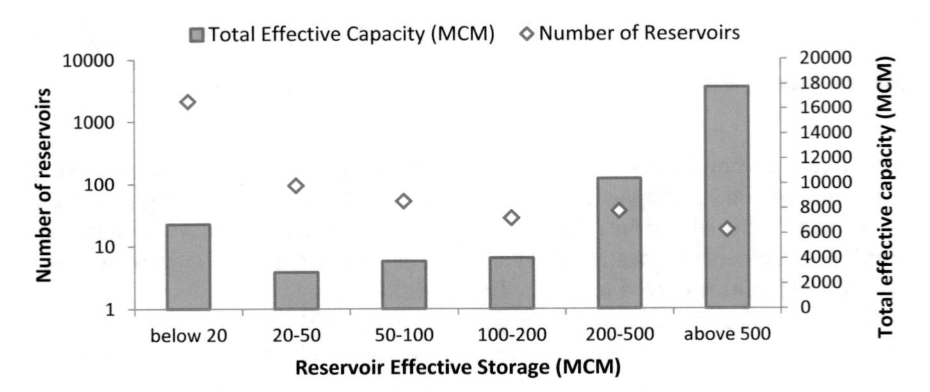

Fig. 6.4 Distribution of reservoirs as per the effective storage. *Source* Authors' analysis using the data from the various water accounting reports of the Government of Maharashtra)

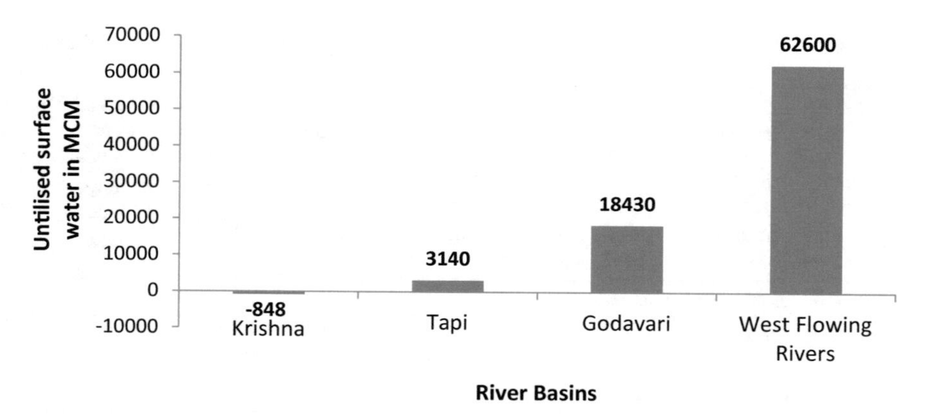

Fig. 6.5 Unutilized surface water in Maharashtra after accounting for all project diversions and lifting. *Source* Authors' estimates using the data from various reports of the Government of Maharashtra

6.3.3 Un-utilized Water Resources in Different River Basins of Maharashtra

As per the results of the analysis presented and discussed in Chap. 5, there is sufficient surface water available in some of the river basins of Maharashtra (refer to Fig. 6.5) to supply to the majority of districts where groundwater-based schemes will not be able to provide dependable water supply for drinking purposes. Some of the sub-basins of Godavari have surplus water that can be harnessed for parts of the Marathwada and Vidarbha regions. While in Marathwada, the surface water can be used to replace the groundwater-based schemes; in Vidarbha, the surface water can be used to augment the existing groundwater-based schemes. Further, the west-flowing rivers have a large amount of surplus water that can be harnessed for transporting to the Marathwada region. The river lifting schemes or small reservoir-based schemes can be promoted in western Maharashtra districts. Since the catchments in these districts contribute to Krishna basin flows and since Krishna River catchments are already over-developed, replacement water will have to be provided to the command areas of Krishna in Maharashtra using imported water from the west-flowing rivers.

6.3.4 Unit Cost of the Interventions

Using the methodology discussed in Sect. 6.2, the cost of augmenting the existing water supply schemes comes out to be INR 5.14/cu m for a project with a life of 25 years and INR 4.5/cu m for a project with a life of 50 years. The estimates are provided in Table 6.1. These costs are considered for Group 2 and Group 3 schemes.

Table 6.1 Estimates of investments needed for creating water infrastructure per cu m of diverted water

Projects	Total Expenditure up to 2006–07 (in INR 10 million)	Annualized cost at 2006–07 prices in INR 10 million (8% discount rate)	Actual irrigated area in 2006–07 (in hundred thousand ha)	Water diverted (cu m) per ha of irrigated area in 2006–07	Annualized cost (INR) per cu m of water diverted for irrigation	
					2006–07 prices	Adjusted to 2016–17 prices
A] Lifetime expectancy of 50 years						
Major and medium	50,049	4091	14.56	10,292	2.73	4.48
B] Lifetime expectancy of 25 years						
Major and Medium	50,049	4689	14.56	10,292	3.13	5.14

Source Authors' estimates using the data from various reports of the Government of Maharashtra

However, these unit cost figures would at best be representative of the schemes that impound and transfer water within the basin, as most of these schemes considered for the analysis are intra-basin water transfer schemes. As analysis of investment data for water resource projects in different parts of the country has indicated, if the projects are in naturally water-scarce regions, which typically involves the transfer of water over long distances, the unit cost goes up considerably (refer to Table 6.2).

Table 6.2 Average investment cost for irrigation schemes in selected Indian states

Region/state	Annualized capital cost in INR hundred thousand (2012–13)	O&M cost in INR hundred thousand (2012–13)	Estimated supply cost (INR/ha of irrigated area)
A] Naturally water-scarce regions			
Tamil Nadu	105,824	52,102	23,291
Madhya Pradesh	320,252	62,275	24,374
Karnataka	397,216	33,777	32,293
Gujarat	595,642	55,918	71,365
Maharashtra	614,569	178,398	73,423
Undivided Andhra	930,017	839,421	105,110
B] Naturally water-rich regions			
West Bengal	8607	24,026	5827
Bihar	81,615	39,947	7988
Uttar Pradesh	97,677	323,415	10,869
Odisha	137,645	47,568	15,724

Source Authors' analysis

Therefore, for schemes that involve inter-basin water transfer (Group 4 schemes), the earlier cost figures should not be considered.

For such schemes, the investment figures for the Sardar Sarovar Narmada Project in Gujarat, which involved long-distance water transfer from the relatively water-rich Narmada River basin in south Gujarat to water-scarce North Gujarat, Saurashtra, and Kachchh regions, were considered. The total investment cost considered was INR 600 billion and that of annual operation and maintenance to be INR 20 billion. Following the methodology discussed in Sect. 6.2, the unit cost of appropriating and diverting and distributing water to different regions (which, along with an extensive canal network, also involved the major lifting of water for the Saurashtra branch canal) worked out to be INR 7.4/m^3.

For schemes in Group 1 category, groundwater will be the source, and these sources were found to be sustainable. Therefore, the data for one of the existing tube well schemes from that cluster with an assumed life span of 12 years was considered. The unit cost was INR 5.0/cu m. This cost, however, does not include the cost of building the infrastructure within the village for the distribution of water. The reason for not considering this component is that the infrastructure already exists in the villages, and the problem is more of the availability of water throughout the year.

6.4 Scale of Investment Required for Improving Rural Water Supply

Following the methodology discussed in Sect. 6.2, the total water demand for all the districts, which included clusters to be catered to by Group 1 schemes, was estimated at 1,497 MCM per annum. The total investment required to improve rural water supply was estimated to be INR 95.5 billion. Figure 6.6 shows the amount of water

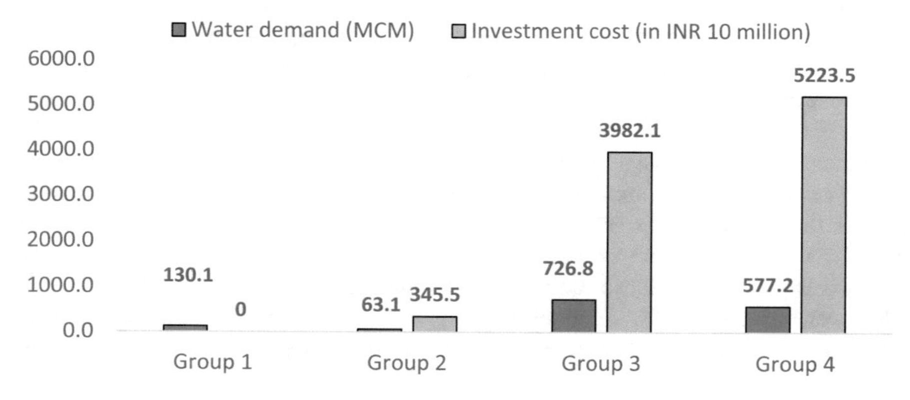

Fig. 6.6 Cluster-wise investment costs for different water supply interventions, Maharashtra. *Source* Authors' estimates

required to meet the rural water demand in different clusters and the amount of investment required. For Group 1 schemes, no investment is considered, as the existing local groundwater-based schemes would be able to meet the estimated requirement of 130.1 MCM per annum.

The issue of the cost-effectiveness of these interventions can be addressed by looking at the extent of dependence on tanker water supply and the amount of money households spend for procuring water. If we go by a conservative estimate of 20% of the rural households (coming under Group 2, Group 3, and Group 4 districts) depending on tanker water for three months in a year on an average, the additional water to be supplied through tankers to overcome water scarcity comes to 77.75 MCM per annum. At an average price of INR 250/m^3 for tanker water (i.e. INR 0.25/litre), the expenditure would be INR 19.43 billion per annum for tanker water supply alone.

6.5 Conclusion

Based on a complex regression model, we had identified the key attributes that would have a positive influence on the sustainability of rural water supply schemes. They are: utilizable aquifer recharge rate; aquifer storage space; extent of gravity irrigation; and limited irrigation demand. Based on how different districts of the state fare on these positive attributes, the overall ranking of the districts is done in terms of the probability of success of groundwater-based drinking water schemes.

Subsequently, the districts were grouped into four different typologies for planning rural water supply schemes, and a strategy was proposed to ensure the sustainability of the water supply in all four of them. The first group of districts is where groundwater-based schemes will be sustainable. The second group is where groundwater-based schemes last for 8–9 months in a year and therefore need to be supplemented by surface water for ensuring a year-round water supply. The third group is where local surface water resources are available in plenty and can be tapped for sustainable rural water supply. The fourth group is where inter-basin water transfer would be required for providing bulk water supply for supporting regional new water supply schemes. It has been assessed that the Godavari River basin and the west-flowing rivers in Maharashtra have a substantial amount of un-utilized surface water that can be used for water transfer to the areas where groundwater-based schemes are unsustainable.

For the surface water schemes based on intra-basin water transfer (Group 3) and for strengthening the existing groundwater-based schemes with surface water for enabling conjunctive use (Group 2), the estimated unit cost will be INR 4.5/m^3. However, the amount of water required per household in Group 2 would only be 1/4th of that required for creating new surface water schemes (Group 3), as the households can tap groundwater from the existing sources for nearly 9 months of the year. Hence, the total investment cost for storage will be just 25% of that considered for Group 3 for the same population. But this water needs to be transported to local storages in smaller quantities throughout the year to reduce the cost of conveyance pipeline. For the bulk transfer of surface water to the regional water supply schemes planned for Group 4, the cost will be INR 7.4/m^3 considering a system life of 50 years. For the groundwater-based schemes supplying year-round water (Group 1), the cost works out to be INR 5/cu m considering a system life of 10 years.

Overall, the total cost of the investment required for supplying 1,497 MCM of water, which can ensure year-round rural domestic water security in Maharashtra, is estimated to be INR 95.5 billion (US$1.25 billion). More importantly, this has been found cost-effective as the cost of ensuring summer water security through supplying 77.75 MCM of water through tankers alone works out to be far greater, i.e. INR 19.43 billion per annum.

References

Biswas-Tortajada A (2014) Gujarat state-wide water supply grid: a step towards water security. Int J Water Resour Dev 30(1):78–90

Chaudhuri S, Roy M, McDonald LM, Emendack Y (2020) Water for all (*Har Ghar Jal*): Rural water supply services (RWSS) in India (2013–2018), challenges and opportunities. Int J Rural Manag 16(2):254–284

Jagadeesan S, Kumar MD (2015) The sardar sarovar project: Assessing economic and social impacts. Sage Publications, New Delhi

Kumar MD (2017) Market analysis: desalinated water for irrigation and domestic use in India, prepared for securing water for food: a grand challenge for development in the Center for Development Innovation, U.S. Global Development Lab. DAI Professional Management Services, US

Kumar MD (2018) Water policy, science and politics: an Indian perspective. Elsevier Science, Amsterdam

Reddy VR (2018) Techno-institutional models for managing water quality in rural areas: case studies from Andhra Pradesh, India. Int J Water Resour Dev 34(1):97–115

Reddy VR, Rammohan Rao MS, Venkataswamy M (2010) 'Slippage': the bane of drinking water and sanitation sector (a study of extent and causes in rural Andhra Pradesh). WASHCost India-CESS Working Papers, Hyderabad, India

Reddy VR, Jayakumar N, Venkataswamy M, Snehalatha M, Batchelor C (2012) Life-cycle costs approach (LCCA) for sustainable water service delivery: a study in rural Andhra Pradesh, India. J Water Sanit Hygiene Dev 2(4):279–290

Chapter 7
Managing Groundwater Quality for Drinking Water Security in India: Emerging Challenge

7.1 Introduction

Groundwater is generally less susceptible to pollution when compared to surface water bodies. Also, the natural impurities in rainwater, which replenishes groundwater systems, get removed while infiltrating through soil strata. But in India, where groundwater is used intensively for irrigation, domestic and industrial purposes, a variety of land and water-based human activities are causing pollution of this precious resource (Kumar and Shah 2006).

Indiscriminate disposal of untreated and partially treated industrial effluents and sewage from cities and towns into rivers and lakes in the past had caused pollution of groundwater in many pockets of India (Kumar and Tortajada 2020). Its over-exploitation is causing aquifer contamination in certain instances, while in certain others, its unscientific development with insufficient knowledge of groundwater flow dynamic and geohydrochemical processes has led to its mineralization (Farooqi et al. 2007; Kumar and Shah 2006; Li et al. 2015). Problems of groundwater quality deterioration have serious implications for water availability for drinking in rural areas in India where nearly 80% of this demand is met from wells (open wells, bore wells, and tube wells) and handpumps.

Though the magnitude of the problem posed by industrial effluents and municipal sewage is slowly reducing over time, with a consistent increase in wastewater treatment capacity in India's cities (Kumar and Tortajada 2020), the millions of toilets now being built in rural areas of the country that provide on-site sanitation have become a major source of concern for their potential to cause widespread pollution of India's groundwater resources with nitrates and microorganisms. Protecting the shallow groundwater from microbial contamination from onsite sanitation facilities in rural areas is proving to be a major institutional challenge for the government agencies concerned with drinking water supplies.

In this chapter, we will first discuss the extent of contamination and pollution of groundwater in India from the point of view of using it for human consumption, particularly the emerging problem of microbial pollution of groundwater due to

M. Dinesh Kumar et al., *Drinking Water Security in Rural India*, Water Resources Development and Management, https://doi.org/10.1007/978-981-16-9198-0_7

on-site sanitation through millions of toilets, based on a synthesis of the available scientific evidence on the subject. We will then discuss the potential health impacts of groundwater contamination and pollution and provide some available evidence to this effect. This is followed by a narration of the issues and challenges in tackling groundwater contamination and pollution for protecting drinking water sources from technical and institutional perspectives. The technologies available for the treatment of poor-quality groundwater to prevent adverse health impacts from consuming it will be discussed in the next section. While doing that, we will introduce the concept of water quality surveillance based on risk assessment as a way to prevent groundwater contamination. In the subsequent section, the emerging challenges with regard to the management of the treatment systems will be discussed, followed by some policy inferences presented in the concluding section.

7.2 Extent of Groundwater Contamination and Pollution and Their Impacts

Poor quality of groundwater in India is mainly due to naturally occurring contaminants and pollutants added through anthropogenic activities. Natural contamination of groundwater is chemical in nature and is mainly due to the concentration of certain chemical constituents above permissible levels as specified by the World Health Organization. As regards the pollution of groundwater, it is chemical and biological.

7.2.1 Chemical Contamination of Groundwater and Its Impacts

Groundwater contamination problems encountered in India that are naturally occurring include: high salinity (in the form of high TDS and high chloride), high fluoride concentration, high nitrate concentration, high arsenic concentration, and high iron. Among all these, salinity is the most extensive contaminant in India's aquifers.

The incidence of fluoride above permissible levels of 1.5 ppm occurs in 20 states and union territories, namely Andhra Pradesh, Assam, Bihar, Chhattisgarh, Delhi, Gujarat, Haryana, Jammu and Kashmir, Jharkhand, Karnataka, Kerala, Madhya Pradesh, Maharashtra, Orissa, Punjab, Telangana, Rajasthan, Tamil Nadu, Uttar Pradesh, and West Bengal affecting a total of 225 districts, according to some estimates (CGWB 2010). Some other estimates find that 65% of India's villages are exposed to fluoride risk (Kumar and Shah 2004).

High levels of salinity are reported from all these states and also the NCT of Delhi, and affects 163 districts (CGWB 2010) and six blocks of Delhi. Iron content above the permissible level of 0.3 ppm is found in 23 districts from 4 states, namely Bihar,

Rajasthan, Tripura and West Bengal, and also coastal Orissa and parts of Agartala valley in Tripura.

High levels of arsenic above the permissible levels of 50 parts per billion (ppb) are found in the alluvial plains of Ganges covering six districts of West Bengal.

High levels of salinity—measured in terms of total dissolved solids (TDS)—renders groundwater unfit for drinking. Per the Bureau of Indian Standards (BIS), the permissible level is 1000 ppm for potable water, though some Indian states have adopted more liberal standards, with water sources of higher salinity (up to 1500 ppm) often used for drinking water when no alternatives are available. There are other sources of chemical contamination of groundwater in India. They are: fluoride, nitrates and arsenic (Kumar 2017). The permissible level of fluoride for drinking water supply is 1.5 ppm. The permissible level of nitrate is 45 ppm. The permissible level of arsenic is 0.05 ppm or 50 parts per billion.

Figure 7.1 shows the geographical extent of salinity in groundwater in India as per the latest data available from the Central Ground Water Board. Figures 7.2, 7.3, and 7.4 show the districts, where groundwater is affected by these contaminants in terms of the percentage of administrative blocks affected, for fluorides, arsenic, and nitrates, respectively. This assessment is based on spot data on the incidence of high levels of these contaminants in groundwater, as obtained from the observation wells in these blocks, as the actual data on the areal extent of contamination are not available (CGWB 2010). These maps do not suggest the degree of contamination within a certain locality and instead, suggest the areal spread of contamination.

These problems of high concentration of chemicals in groundwater pose significant challenges for water supply and public health professionals, one major reason being that the concentration of chemical contaminants is below permissible levels while the source water is tested at the time of commissioning, but the same increases over a period of time, as the abstraction of water from the aquifer for various uses increases. Since continuous monitoring of water quality at the field level is absent, ensuring the safety of the supplied water over long time periods is difficult.

The second reason is that the presence of chemical contaminants such as fluoride, arsenic, and nitrates in water cannot be detected without the help of water quality testing equipment. High fluoride content is often detected from such symptoms on human beings as yellowing of teeth, damaged joints and bone deformities, which occur from long years of exposure to fluoride-containing water. Due to this reason, by the time the community realizes the 'menace', a large section of the population is already affected. A survey by the International Water Management Institute (IWMI) in north Gujarat in 2003 showed 42% of the people covered in the sample survey (28,425) were affected; while 25.7% were affected by dental fluorosis, 6.2% were affected by muscular skeletal fluorosis, and 10% by both (Kumar and Shah 2006).

The potential biological and toxicological effects of using fluoride-contaminated water are also dangerous. A study on fluorotic populations of north Gujarat revealed an increase in the frequency of sister chromatic exchanges in fluorotic individuals indicating that fluoride might have a genotoxic effect (Sheth et al. 1994). Fluoride had been reported to cause depressions in DNA and RNA synthesis in cultured cells. Another study on the effects of fluorides in mice showed significant reductions

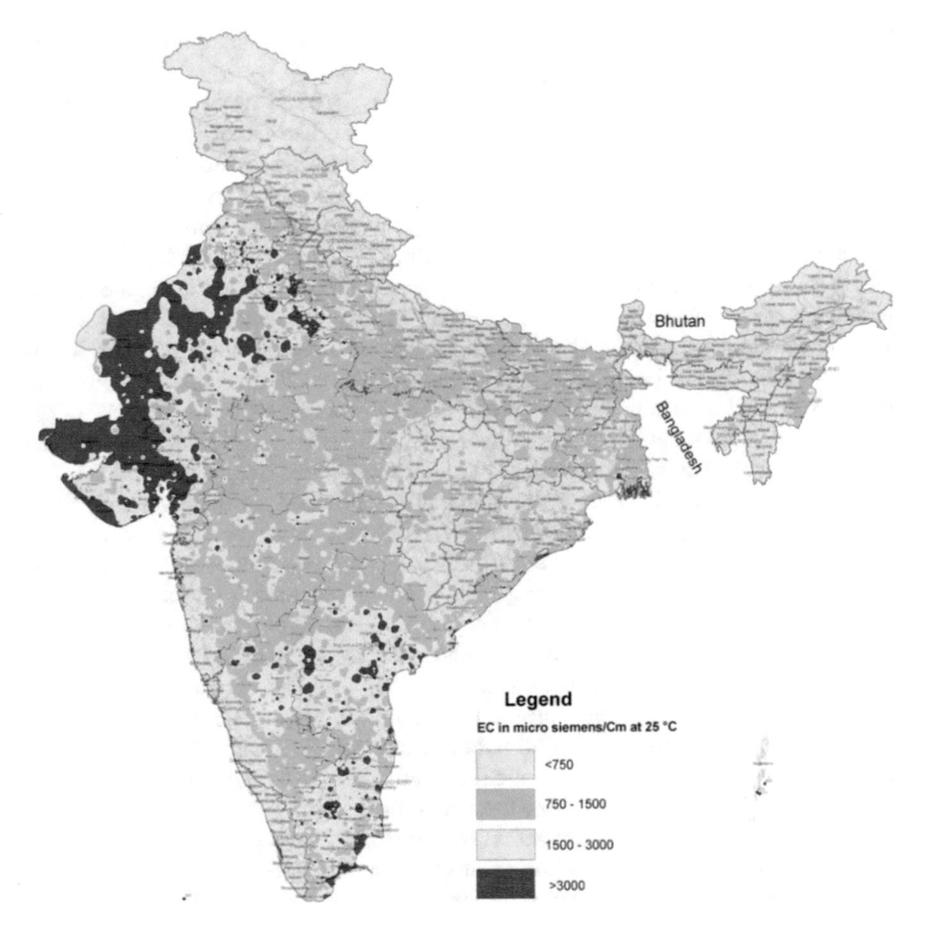

Fig. 7.1 Map showing groundwater salinity levels in India. *Source* CGWB (2010)

in DNA and RNA levels (Jha et al. 2012). Conditions including ageing, cancer, and arteriosclerosis are associated with DNA damage and its disrepair. Prolonged exposure to water containing salts (TDS above 500 ppm) can cause kidney stones, a phenomenon widely reported in north and coastal Gujarat.

Arsenic contamination of drinking water causes a disease called arsenicosis, for which there is no effective treatment, though consumption of arsenic-free water could help affected people at early stages of ailment to get rid of the symptoms of arsenic toxicity. Arsenic contamination is by far the biggest mass poisoning case in the world, putting 20 million people from West Bengal and Bangladesh at risk, though some other estimates put the figure at 36 million people (Kumar and Shah 2006).

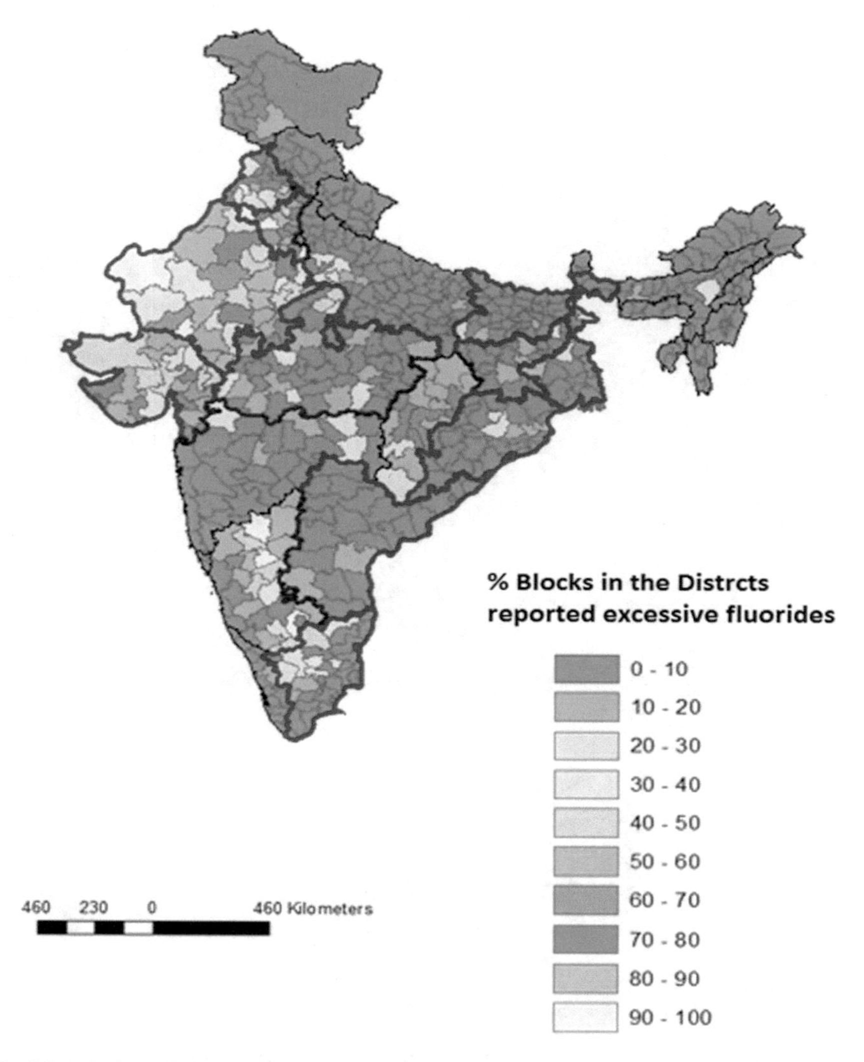

Fig. 7.2 Districts affected by fluoride contamination in groundwater

7.2.2 Groundwater Pollution and Its Impacts

In India, pollution of groundwater occurs from several sources. They are mainly: (i) untreated and partially treated effluent from industries in urban and rural areas, (ii) untreated or partially treated municipal sewage and effluent from septic tanks in urban areas (barring those urban areas where the sewage is fully treated); (iii) entirely untreated rural domestic wastewater; (iv) effluent from septic tanks and toilet pits in

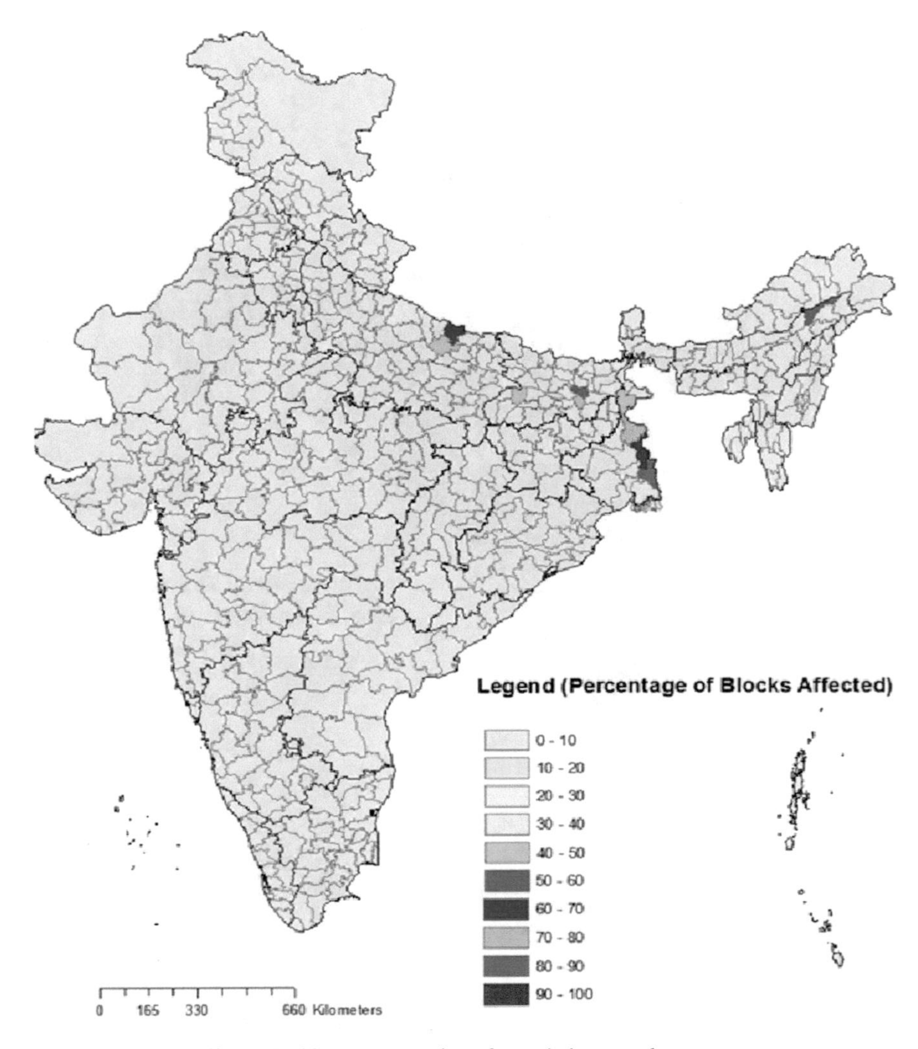

Fig. 7.3 Districts affected by high concentration of arsenic in groundwater

rural areas; (v) runoff and percolation of excess water from irrigated fields containing fertilizer and pesticide residues.

The type of pollution the above-mentioned sources can cause to groundwater are either chemical or biochemical. The pollution from industrial effluents is mainly chemical—salinity, heavy metals, pesticides, and high BOD and COD. The type of pollution caused by domestic effluent can be chemical (due to the use of soaps, detergents, acids, micro plastics, etc.), biochemical (food waste, garbage, paper, and effluent from washrooms and toilets), and bacteriological (total coliform count, faecal coliform, and *Escherichia coli*, originating from blackwater). The pollution

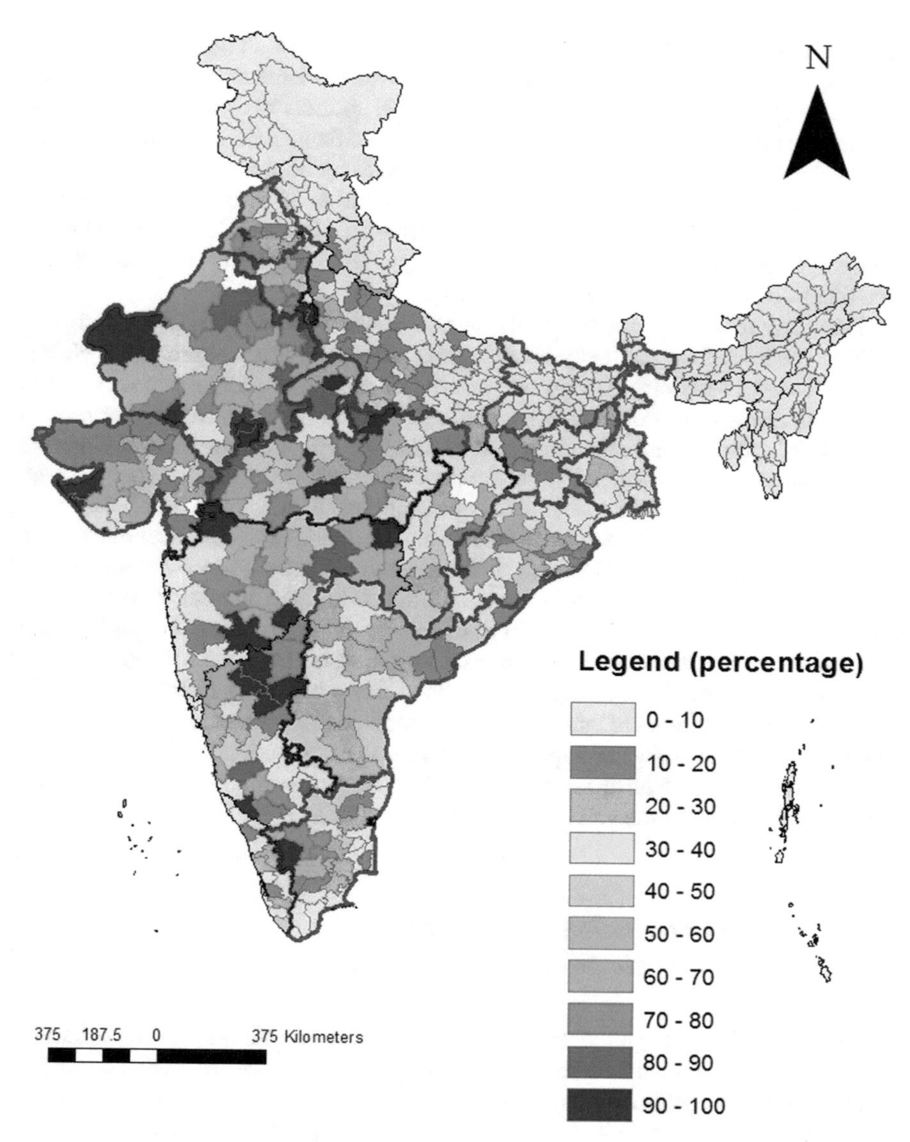

Fig. 7.4 Districts affected by high concentration of nitrates in groundwater in India

caused by runoff and percolation from agricultural fields is chemical (with nitrates and phosphates, and some complex compounds originating from pesticides).

Pollution of groundwater due to industrial effluents and municipal waste in water bodies is a major concern in many cities and industrial clusters in India. A survey undertaken by the Centre for Science and Environment from eight places in Gujarat, Andhra Pradesh, and Haryana reported traces of heavy metals such as lead, cadmium, zinc, and mercury. The shallow aquifer in Ludhiana city, the only source of its drinking water, is polluted by a stream that receives effluents from 1300 industries. Excessive withdrawal of groundwater from coastal aquifers has led to induced pollution in the form of seawater intrusion in Kachchh and Saurashtra in Gujarat, Chennai in Tamil Nadu, and Calicut in Kerala (Kumar and Shah 2006).

Presence of heavy metals in groundwater is found in 40 districts from 13 states, viz. Andhra Pradesh, Assam, Bihar, Haryana, Himachal Pradesh, Karnataka, Madhya Pradesh, Orissa, Punjab, Rajasthan, Tamil Nadu, Uttar Pradesh, West Bengal, and five blocks of Delhi (Kumar and Shah 2004).

There are very few estimates of the public health consequences of groundwater pollution as it involves methodological complexities and logistical problems. Nevertheless, levels of toxicity depend on the type of pollutant. Mercury is reported to cause impairment of brain functions, neurological disorders, retardation of growth in children, abortion, and disruption of the endocrine system, whereas pesticides are toxic or carcinogenic. Generally, pesticides damage the liver and nervous system. Tumour formation in the liver has also been reported (Kumar and Shah 2006).

Non-point pollution caused by fertilizers and pesticides used in agriculture, often dispersed over large areas, is a great threat to fresh groundwater ecosystems. Intensive use of chemical fertilizers in farms and indiscriminate disposal of human and animal waste on land result in leaching of the residual nitrate causing high nitrate concentrations in groundwater. Nitrate concentration is above the permissible level of 45 ppm in 11 states, covering 95 districts and two blocks of Delhi. DDT, BHC, carbamate, Endosulfan, etc. are the most common pesticides used in India. But, the vulnerability of groundwater to pesticide and fertilizer pollution is governed by soil texture, the pattern of fertilizer and pesticide use, their degradation products, and total organic matter in the soil (Kumar and Shah 2006).

7.2.3 On-Site Sanitation: An Emerging Source of Large-Scale Groundwater Pollution

An emerging source of pollution threat to groundwater in India is on-site sanitation. While millions of toilets with septic tanks and pit-type latrines are being built in the rural areas for disposal of human waste under the Swatch Bharat Mission (SBM) to make the country open defecation free (ODF), its impact on groundwater, unfortunately, has not received the attention it deserves. The partially treated effluent from septic tanks and toilet pits can be a major source of groundwater pollution. The effluent from septic tanks is partially treated and has high BOD, as several research studies from different parts of the world suggest. Microbial and nitrate contamination of groundwater from pit latrines have been widely reported (Banerjee 2011; Pujari et al. 2012; Verheyen et al. 2009; Zingoni et al. 2005).

The impacts of both the contaminants on human health are serious. The health impact of bacteriological contamination of drinking water will be immediate on humans when they are exposed to it, in the form of diarrhoea. In the case of nitrates, however, only prolonged exposure will cause the disease. Consuming too much nitrate can affect the way the blood carries oxygen and can cause 'methemoglobinemia'. Babies under the age of six months are at the highest risk of developing methemoglobinemia. This illness can cause skin to turn a bluish colour and can result in serious illness or death. Other symptoms connected to methemoglobinemia include decreased blood pressure, increased heart rate, headaches, stomach cramps, and vomiting (ATSDR 2015).

Contamination takes place in the event of a pathway existing between the on-site sanitation system and groundwater body. Groundwater pollution due to on-site sanitation systems has been dealt with by many workers (Gerba and Bitton 1984; Hagedon 1984; Chidavaenzi et al. 2000).

Concern about groundwater pollution due to on-site sanitation systems relates primarily to unconfined and, to a lesser degree, to semi-confined aquifers. If groundwater supplies are drawn from deep and confined aquifers, on-site sanitation does not pose a significant hazard. Recent studies by Lawrence et al. (2001) highlight the role of hydrogeology in determining the degree of contamination of groundwater from on-site sanitation. Studies carried out in the USA on groundwater pollution from septic tank effluent have been a major source of information. However, the effect of the difference in the design and construction of septic tank disposal systems, and the proposed sanitation systems may be significant. The studies carried out were in Columbia in sand-clayey sand (Kligler 1921), in Alabama in fine-sand medium (Caldwell 1938). Besides, studies were conducted in sandy clay in Texas (Brown et al. 1979) and fractured rocks in Colorado (Allen and Morrison 1973).

The majority of the field studies were confined, mainly to fine-grained sediments, which are of low risk, and consequently most suitable for on-site sanitation. There is a need to obtain more information on other soil types. There is a need for a classification of hydrogeological environments in relation to pollution risk. This would be of great value in the appraisal and implementation of on-site sanitation schemes. The factors affecting the pollution of groundwater from on-site sanitation are well documented and are as follows: depth to water table; hydraulic loading; the structure and texture of soils in the unsaturated zone; and presence of fissures in the case of hard rock formations. After Gerba et al. (1975), the factors affecting microbial contamination of groundwater from on-site sanitation can be summarized as follows:

- The chances of contamination increase significantly in geological settings where the water table is very shallow (1–15 m).
- The unsaturated zone represents the first line of defence against aquifer pollution. Soil provides a very effective natural treatment system. It has the ability to remove faecal microorganisms and chemical/biochemical compounds. The nature of the geological strata and thickness of the unsaturated zone determine the risk of pollution. While natural flow rates in the unsaturated zone of almost all formations do not normally exceed 0.3 m/day (Lawrence et al. 2001), it can be more than an order of magnitude higher in the case of fractured formations. Flow rates in excess of 5 m/day may occur in fissured rocks and coarse gravel (Franceys et al.

1992) and the potential for groundwater contamination under these conditions is extremely high. Thus, rock type, especially the grade of consolidation and presence of fractures, are key factors in assessing the vulnerability of aquifers to pollution.

- The key factor in reducing microbiological contamination of groundwater is the maximization of effluent residence time in the unsaturated zone. Many contaminants, especially microorganisms, are rendered harmless or reduced to low concentrations by natural processes when the movement of the contaminants in the sub-surface is slow. The natural treatment processes, such as filtration, are more efficient in fine-grained unstructured soils. Structures such as root channels, animal burrows, natural voids and fissures commonly lead to short-circuiting of the unsaturated zone with a consequent reduction in the residence time and natural treatment. This may lead to a greater risk of groundwater pollution.
- Clogging of the filtration surface in the latrine pit enhances bacteria and virus removal processes so that the risk of pollution from microorganisms diminishes after the first 100 days or so of pit usage. But, it can reduce the infiltration of the effluent, thereby affecting the capacity to reduce the BOD, COD, nitrate, etc.
- More specifically, the risk of microbiological groundwater pollution will be minimal where more than 2 m of fine unsaturated soils are present beneath the latrine pit, provided the hydraulic loading in the pit does not exceed 50 mm/day.
- In the saturated zone, pollutants move with the groundwater causing a pollution plume to develop from the pollution source. Contamination removal processes take place in the saturated zone but at a lesser rate compared to the unsaturated zone since groundwater moves more rapidly. Within the saturated zone, dispersion and dilution play an important role in reducing the concentration of the contaminants.

Improper design, construction, operation, or maintenance of on-site sanitation systems can lead to failure due to the loss of infiltration capacity, with the consequent surfacing of effluent. Such failures are quite frequently reported. In well-designed septic tanks, the solid matter does not represent a significant hazard, but the soak pit causes both microbiological and chemical contamination. There is a potential threat to groundwater where hydraulic loads are high, and they exceed natural attenuation potential in the sub-surface. However, an equally important and more insidious failure is that of inadequate effluent purification.

The organic matter gets filtered while passing through the soil formations. They also get adsorbed and digested through aerobic and anaerobic processes. The microorganisms such as bacteria, viruses, and fungi get adsorbed.

Adsorption of both viruses and bacteria is highest in soils with high clay content and is favoured by a long residence time—that is, when flow rates of effluent are slow. Since the flow is much slower in the unsaturated zone than in the saturated zone, the contact time is longer between soil and effluent and thereby increase the chances of adsorption. Adsorbed microorganisms can be dislodged, for example, by flushes of effluent or following heavy rainfall, and may then pass into lower strata of the soil. Both viruses and bacteria live longer in moist conditions than in dry conditions.

Bacteria live longer in alkaline soils than in acidic soils. The bacteria also survive well in soils containing organic material, where there may be some regeneration.

The survival of bacteria and viruses in soils also depends on the temperature. The reduction of polioviruses held for 84 days in loamy sand was less than 90% at 4 °C, but 99.99% at 20 °C. It was also found that aerobic inactivation was more rapid under non-sterile versus sterile conditions and that anaerobic conditions led to a reduction in inactivation. In a nutshell, while the formation conditions influence the removal process, the survival of these microorganisms in the formations depends on the temperature, pH value, moisture content, etc. But, the dominant factors for bacteria survival are temperature and moisture (Gerba et al. 1975).

Nitrate has been the most widely investigated chemical contaminant, because of its use as an indicator of faecal contamination and its impact on human health (Graham and Polizzotto 2013). Although a number of studies that detected total or faecal coliforms in wells did not see elevated nitrate concentrations (Dzwairo et al. 2006; Howard et al. 2002; Still and Nash 2002), other studies have reported nitrate concentrations >100 mg/L (Banks et al. 2002; Mafa 2003; Pujari et al. 2012). Nitrate concentrations in groundwater near latrines were frequently higher than the levels in the local area (Baars 1957; Caldwell and Parr 1937; Chidavaenzi et al. 2000; Zingoni et al. 2005).

However, WHO (2006) notes that in spite of growing evidence of an association between nitrate concentration in groundwater and the proximity to latrines, pinpointing the actual sources of nitrate in groundwater has proved challenging. This is because nitrate may be derived from numerous potential sources in urban and rural environments, including latrines, plant debris, animal manure, garbage repositories, livestock pens, soil, and fertilizers (Girard and Hillaire-Marcel 1997; Howard et al. 2002; Vinger et al. 2012); and nitrate can be formed and lost through natural soil processes (Jacks et al. 1999).

Girard and Hillaire-Marcel (1997) used nitrogen isotopes to determine the source of nitrate pollution in a fractured rock aquifer of Niger. Nitrate concentrations in wells reached 11.6 milliequivalents/L, which according to them may have been a consequence of contamination by latrines and deforestation (Girard and Hillaire-Marcel 1997).

A more common approach in establishing the causality is to compare areas with similar environmental characteristics but different population and latrine densities. By analysing water samples from installed boreholes in an informal settlement in Zimbabwe, Zingoni et al. (2005) showed that the highest nitrate concentrations in groundwater (20–30 mg/L) were associated with the highest population and density of pit latrines in the settlement. In Siberia and Kosova, nitrate concentrations were sometimes >100 mg/L in groundwater of villages with high latrine densities and minimal septic tanks, but concentrations were below hazardous levels in agricultural and unpopulated settings (Banks et al. 2002). In Senegal and South Africa, groundwater nitrate concentrations have also been correlated with proximity to pollution sources (Tandia et al. 1999; Vinger et al. 2012).

Like in the case of microbial contamination, environmental factors play a role in governing nitrate pollution of groundwater: from latrines. Pujari et al. (2012)

compared the impacts of on-site sanitation in two Indian megacities and concluded that hydrogeological conditions were strong predictors of the threat of nitrate contamination of well water; an area with shallow water table was more susceptible to pollution than an area with a deeper water table. Lewis et al. (1982) found that in eastern Botswana, the build-up of nitrogenous latrine effluent in soils and its subsequent downward leaching promoted dissolved nitrate concentrations: >500 mg/L in groundwater, and concluded that the fissured aquifer allowed rapid contaminant transport. Whereas soil type immediately below the pit is likely to influence the degree of nitrate transport (Caldwell and Parr 1937), associations with soil type have not always been observed (Nichols et al. 1983).

Hence, it can be summarized that soil, hydrological, climatic and geohydrological factors, and the effluent load will have a strong bearing on microbial and biochemical contamination of groundwater. Having said that, what is important is that the condition vis-à-vis these factors varies from region to region and place to place in India. Keeping in view the conditions with regard to these factors, the susceptibility of each region to pollution should be examined.

7.3 Issues in Tackling Groundwater Contamination and Pollution

The first step towards evolving measures to prevent and cure groundwater quality deterioration is generating reliable and accurate information through water quality monitoring (WQM) to understand the actual source/cause, type and level of contamination. However, there are a few observation stations in the country that cover all the essential parameters for water quality and hence the data obtained are not decisive on the water quality status. Secondly, WQM involves expensive and sophisticated equipment that are difficult to operate and maintain and require substantial expertise in collecting, analysing and managing data. Since water technology is still not advanced in India, it is very likely that the available data is less reliable. The existing methodology for WQM is inadequate to identify the various sources of pollution. Integration of data on water quality with data on water supplies, which is very important from the point of view of assessing water availability for meeting various social, economic, and environmental objectives, is hardly done. And finally, in the absence of any stringent norms on water quality testing, results can change across agencies depending on sampling procedure, time of testing, and testing instruments and procedure.

Now let us examine technical issues in mitigating contamination. For seawater intrusion, artificial recharge techniques are available in India for different geohydrological settings. Artificial recharge could push seawater–freshwater interface seawards. These techniques can also be used to reduce the levels of fluoride, arsenic, or salinity in aquifer waters on the principle of dilution. But the issue is of availability of good water for recharging in arid and semi-arid regions, given the large aerial extent of contaminated aquifers. For industrial pollution, the issues are of three types: pumping out polluted water from the aquifer; treated this water to safe limits, and replenishing the depleted aquifer with freshwater. Technically, feasible methods to clean polluted water often don't exist due to highly toxic substances in trade effluents, as seen in a case in Rajasthan where a sulfuric acid manufacturing unit rendered drinking water sources in 22 villages useless. Finding enough freshwater for replenishment was also a problem there.

In the Indian context, it is not economically viable to clean aquifers. In the case of arsenic, methods for in situ treatment have already been in use in developed countries. In the USA, zero-valent, iron permeable reactive barriers (PRBs) are used in situ to remove chromium and several chlorinated solvents in groundwater and are tested successfully for removing arsenic. India is too poor to afford some of the technologies that are successfully tried out in the West, especially the USA because they are prohibitively expensive. The cost of cleaning the aquifer in the Rajasthan case was estimated to be Rs. 40 crores.

In India, groundwater quality monitoring is primarily the concern of the Central Ground Water Board and state groundwater agencies, where each of them sets up their monitoring network (Kumar and Shah 2006). But there are issues concerning the adequacy of scientific data available from them:

- The network of monitoring stations is not dense enough.
- Water quality analysis excludes critical parameters that help detect pollution by fertilizer and pesticide, heavy metals, and other toxic effluents.
- The available scientific data, particularly that on pollution is of civil society institutions, and there is a paucity of such institutions that are capable of carrying out such professionally challenging, technologically sophisticated, and often politically sensitive tasks.
- There is a very poor network of observation stations in rural areas of the country for monitoring groundwater quality, especially for microbial contamination, which can become rampant with the widespread use of pit latrines.

The Central Pollution Control Board (CPCB) and the State Pollution Control Boards (SPCBs) are the pollution watchdogs in India (Kumar and Shah 2006). As of 2012, India had 2500 water quality monitoring stations spread over 28 states and six union territories, covering rivers, lakes, tanks, ponds, creeks, and seawater channels, canals, drains, water treatment plants (raw water) and groundwater. Of the 2,500 stations, 807 are for monitoring groundwater[1] (CPCB 2012). The inland water quality monitoring network is operated under a three-tier programme: the Global Environmental Monitoring System, the Monitoring of Indian National Aquatic Resources System, and the Yamuna Action Plan (CPCB 2012). In addition to general parameters, and core parameters such as pH, dissolved oxygen (DO), and biochemical oxygen demand (BOD), biological monitoring and monitoring of trace metals and

[1] The rest are for monitoring rivers (1275 nos.), lakes (190 nos.), drains (45 nos.), canals (41 nos.), tanks (12 nos.), creeks and seawater channels (41 nos.), ponds (79 nos.), and water treatment plants (raw water) (10 nos.) (CPCB 2012).

pesticides are undertaken (CPCB 2012). The monitoring system of groundwater quality, however, is not specifically designed to cover 'non-point' pollution from agriculture (Kumar and Shah 2004) and on-site sanitation.

An analysis of the performance of the Gujarat State Pollution Control Board (GPCB) in the Sabarmati River Basin showed that of the four priority areas identified by the Board for operations, its performance has been satisfactory in only identification of areas facing severe pollution. The monitoring ability itself was doubtful as the agency maintains only two observation wells for groundwater quality monitoring in the entire basin. The GPCB also lacks adequate staff to carry out its functions (Kumar and Shah 2006).

There are problems associated with the institutional design itself. The SPCBs perform the dual functions of monitoring pollution and enforcing pollution control norms. But, the fact that regular WQM and its proper dissemination itself could question the existence of the Boards as an enforcement agency creates a disincentive for them to perform the first function meaningfully. Also, the agency lacks legal teeth and administrative apparatus to penalize polluters. This reduces the effectiveness of the agency in enforcing pollution control norms. The fact that the cost of pollution is much less than the cost of treatment works as a disincentive for polluters, whereas the Boards are not mandated to execute environmental management projects (Kumar and Shah 2006; Kumar and Tortajada 2020).

Groundwater contamination most often occurs due to geohydrochemical processes activated by pumping (Farooqi et al. 2007; Li et al. 2015). Once contamination starts, very little can be done to check it except a total ban on pumping. But this is very difficult, as millions of rural families in India depend on groundwater for sustaining irrigated agriculture and livelihoods. Any legal/regulatory interventions to ban pumping would mean depriving communities of their traditional rights. Though *de jure* rights in groundwater are not clear, landowners enjoy the de facto right to extract groundwater under their land. While nitrate pollution can be properly controlled by following the recommended dosage of fertilizers, crop rotation, proper timing of fertilizer application, and use of organic manure instead of chemical fertilizers, there are no institutional regimes governing fertilizer use and dumping of animal waste (Kumar and Shah 2006).

As regards pollution of groundwater from on-site sanitation and agricultural practices, it is highly dispersed unlike that from industrial effluents. There are millions of toilets in rural areas now, and they are not scientifically designed. Agricultural operations involving irrigation and chemical fertilizer extend over very large areas covering millions of ha of land. There are big technological challenges in removing pollutants from fertilizer and pesticide residues and human waste from the aquifers. That issue notwithstanding, groundwater pollution from on-site sanitation and agricultural practices depends on several complex environmental and technical factors. There is no proper information available on the extent of source pollution by these sources. With imperfect information about the benefits of taking such cleaning measures, the economic viability of the treatment measures can come under serious questioning.

As regards measures to prevent pollution, on the agriculture side, it is extremely difficult to change agricultural practices that would reduce the nitrate load in irrigation return flows, as such measures may have a significant impact on crop yields and consequently on farmer income. India also does not have any regulations on the use of nitrates in crop production and disposal of animal waste on land, two potential sources of nitrate pollution of groundwater. As regards sanitation, designing septic tanks and pit-type latrines to suit the site condition or siting them to avoid groundwater pollution are technically complex challenges as they involve several considerations—such as soil hydraulic properties, fractures in the underlying rock formations, depth to water table, seasonal water level fluctuations, rate of loading of the effluent, etc.

7.4 Reducing Adverse Impacts on Human Health

In areas where freshwater sources are not available for the provision of drinking water, to reduce the adverse impacts of consuming poor-quality water, water treatment options need to be explored. However, such options are preferred when three conditions are satisfied: (i) a large area is affected by water quality problems; (ii) the quality of the source water does not vary widely over time, and (iii) the concentration of the contaminant by and large remains more or less the same and above permissible levels. Generally, this is the case of chemical contaminants in groundwater such as fluoride, salinity, and arsenic. In such cases, depending on the type of contaminants in drinking water, the treatment option can be chosen.

Reverse osmosis (RO) is a process to get rid of all the impurities in drinking water, including deadly ions and organisms and pesticide/fertilizer residues. Under RO systems, water is made to pass through a membrane having a pore size of 0.0001 μm under high pressure. Only 5–10% of the ions are able to slip through the membrane, which is well within acceptable levels as per all standards including WHO, BIS, etc. RO systems are suitable for removing several of the toxic substances present in water in dissolved form, including fluoride, fertilizer and pesticide residues, and heavy metals. But costs vary, depending on the plant capacity and level of utilization, the level of salinity and other impurities in the water, and the distance from the source of water. Costs can range from Rs. 0.03/litre (for brackish water) to Rs. 0.10/litre (for seawater) (Kumar and Shah 2006).

A household arsenic treatment method is the ferric chloride coagulation system. This involves precipitation of arsenic by adding a packet of coagulant in 25 L of tube well water, and subsequent filtration of the water through a sand filter. Field experiments showed that the arsenic concentration in treated water was nearly 1/20th that of raw water. The cost of the chemical (ferric chloride) for treatment is Rs. 0.09 per litre of raw water to be treated.

Another method for removing arsenic is based on 'sorptive filtration that uses iron coated sand bed'. Water is first put in a bucket and stirred for some time to accelerate the precipitation of excess iron. It is then allowed to pass through a sand filter where the excess iron is filtered out. Finally, the water is passed through an iron coated sand filter. But, the efficiency of removing arsenic reduces drastically beyond a certain bed volume with the arsenic concentration of treated water crossing the permissible limit of 50 ppb. The third method involves filtration of arsenic from raw water by passing it through a gravel media containing iron sludge. An evaluative study showed the first two systems to be superior, with the first one found to be most acceptable to the villagers.

As regards microbial and nitrate pollution of groundwater that can result from poorly designed onsite sanitation systems and agricultural practices, the only option to protect the resources is to frequently monitor the quality of water from drinking water wells using a dense water quality monitoring network in rural areas, using a sound surveillance system developed on the basis of pollution risk assessment. Ideally, water quality surveillance should be taken up in areas where the susceptibility of groundwater to pollution and the public health risk associated with them are more. However, such surveillance systems hardly exist in most Indian states. We will discuss the concepts and practices in water quality surveillance in the next chapter.

Nevertheless, for the removal of microorganisms, disinfectants such as chlorine tablets and bleaching power can be used in the pumped water in prescribed quantities before it is supplied through the distribution system as a precautionary measure. This is the practice currently followed in Indian villages and is being done by the water supply and sanitation committees of the Gram Panchayats.

7.5 Emerging Challenges

The available treatment systems to remove chemical contaminants in groundwater work based on principles in physics and chemistry. Hence, their efficiency depends heavily on maintaining certain specified operating conditions. This would call upon qualified technical manpower for system operations, and regular operation and maintenance, which are mostly absent. As reported by Kumar and Shah (2006), the Gujarat Water Supply and Sewerage Board has set up 28 desalination systems since 1989. All of them became dysfunctional within a very short span of time. Most of the 117 desalination plants commissioned in eight states by government agencies became non-operational due to lack of technical manpower for maintenance and improper selection of membrane.

Most of the treatment systems for drinking water have to be tried out at the community level to be cost-effective and affordable. With no major revenues being accrued by government agencies from domestic water supply services, any additional investment for the provision of safe water for drinking and cooking purposes would induce an unprecedented financial burden. Therefore, it will be more appropriate to build and operate water treatment systems on the principle of full cost recovery. The water supplied from the system has to be affordable to all classes of the society as drinking is essential for survival. Therefore, the unit cost of production should be minimized for commercial viability.

The unit cost of production could be brought down considerably by running the plant at peak capacity, which means creating sufficient demand. Higher demand means a lower selling price of water for commercial viability. An increase in plant capacity can reduce unit cost but this lowers the chances of running the plant at full capacity, which means higher operating costs. Hence, optimal plant design, proper selection of membrane, generating sufficient demand, etc. are important for bringing down the cost of production. The level of professional inputs that go into the management of public water supply systems would be far less than adequate to manage these systems. Over and above, operating costs for agency-run systems are likely to be high due to high administrative overheads. As a result, new techno-institutional models need to be evolved to manage the system in order to make them self-sustaining. Involving the private sector in the provision of clean and safe drinking water would be a major step towards achieving this.

It is ordinary people who raise the alarm about poor water quality. Civil society institutions need to be strengthened to respond to water quality problems quickly. This is possible through imparting better knowledge and information about the nature of groundwater contamination, potential sources of threats to groundwater quality in their region and degrees of vulnerability, the ill-effects of using contaminated water, and the possible preventive measures.

They can in turn put pressure on the line agencies to perform. Strengthening civil society institutions is particularly important because groundwater quality variations in nature are often sporadic; it is extremely difficult for monitoring agencies to establish an elaborate network of water quality monitoring stations especially for complex chemical contaminants such as fluorides, arsenic, nitrates, and iron due to the high costs and technical manpower involved, though for detecting microbial contamination of the water supplies from the system, testing can be undertaken by the local community institution using test kits that are available.

Given the absence of perfect information about the quality of water in various sources, it is also not possible for line agencies to identify appropriate treatment measures. Also, the willingness of people to pay for water is directly linked to their knowledge and awareness about the ill-effects of drinking contaminated/polluted water. Credible and technically competent NGOs can play a big role in strengthening civil society, by generating the vital database on groundwater quality.

7.6 Conclusions and Policy Inferences

Widespread problems of groundwater quality deterioration with the increased presence of chemical contaminants and the growing threat of microbial contaminants will pose significant challenges for safe drinking water in rural India, as groundwater continues to be the major source of rural water supply in the country. The reported adverse health impacts of continued exposure to poor-quality groundwater are many and widespread. Preventive and curative measures against pollution and contamination of groundwater may continue to receive low priority for years to come, due to the institutional inadequacies with respect to monitoring and preventing groundwater pollution and cleaning up the polluted aquifers. At the same time, technological measures to prevent the ill-effects of exposure to contaminated water on human health have received priority in the past (Kumar and Shah 2006; Kumar 2017). Demineralization using an RO system can remove all hazardous impurities from drinking water and would be cost-effective in many situations, where TDS, nitrate, and fluoride in groundwater are above permissible levels. Such situations are encountered in large areas of western Rajasthan.

The cost of demineralization is falling rapidly. Saudi technologists believe desalination costs would fall so rapidly over the coming decades that desalination will be cheaper than pumping coastal aquifers (Kumar and Shah 2006). According to a recent report, the country's desalination capacity has gone up from 9 million cubic metres per day in 2006 to 16.4 million cubic metres per day in 2016. Further, nearly 70% of Saudi Arabia's urban water demands are currently being met from desalinated water. Low-cost treatment methods are available for the removal of arsenic from groundwater (Kumar and Shah 2006),

There are, however, significant challenges that water utilities would face such as: (i) building technical and managerial skills to design, install, operate, and manage water treatment systems; (ii) making people pay for treated water; and, (iii) building knowledge and awareness among communities about groundwater quality issues and treatment measures so as to create demand and willingness to pay for good-quality water. In the long term, policies need to be focused on: (i) building scientific capabilities of line agencies concerned with WQM, water supplies, and pollution control; and (ii) restructuring them to perform WQM and enforcement of pollution control norms effectively and to enable them to implement environmental management projects.

The emerging problem of microbial and nitrate pollution of groundwater from the several millions of latrines being constructed and used in rural areas of the country, apart from the non-point pollution caused by intensive use of chemical fertilizers and pesticides in agriculture, poses a new institutional challenge for the government agencies concerned with rural water supply. Establishing a network for monitoring groundwater resources for the water quality parameters to ensure the quality of the source water will not be viable as the number of such observation wells has to be very large and the frequency of monitoring will have to be very high. Hence, a system of surveillance of water quality for the publicly managed drinking water wells in rural areas based on risk assessment will have to be established.

References

Agency for Toxic Substances and Disease Registry (ATSDR) (2015) ToxFAQs for Nitrate and Nitrite. https://wwwn.cdc.gov/TSP/ToxFAQs/ToxFAQsLanding.aspx. Accessed on 17 Aug 2021

Allen MJ, Morrison SM (1973) Bacterial movement through fractured bedrock. Groundwater 11(2):6–10

Baars JK (1957) Travel of pollution, and purification en route, in sandy soils. Bull World Health Org 16:727–747

Banerjee G (2011) Underground pollution travel from leach pits of on-site sanitation facilities: a case study. Clean Technol Environ Policy 13(3):489–497

Banks D, Karnachuk OV, Parnachev VP, Holden W, Frengstad B (2002) Groundwater contamination from rural pit latrines: Examples from Siberia and Kosova. J Chart Inst Water Environ Manage 16(2):147–152

Brown KW, Wolf HW, Donnelly KC, Slowey JF (1979) The movement of faecal coliforms and coliphages below septic lines. J Environ Qual 8(1):121–125

Caldwell EL (1938) Pollution flow from a pit latrine when permeable soils of considerable depth exist below the pit. J Infect Dis 62:225–258

Caldwell EL, Parr LW (1937) Ground water pollution and the bored hole latrine. J Infect Dis 61(2):148–183

Central Ground Water Board (2010) Ground water quality in shallow aquifers of India. Ministry of Water Resources, Government of India. http://cgwb.gov.in/WQ/gw_quality_in_shallow_aquifers.pdf. Accessed on 20 Aug 2021

Central Pollution Control Board (2012) Status of Water Quality in India. Ministry of Environment and Forests

Chidavaenzi K, Bradley M, Jere M, Nhandara C (2000) Pit latrine effluent infiltration into groundwater: the Epworth case study. Water Sanitation and Health, IWA, London

Dzwairo B, Hoko Z, Love D, Guzha E (2006) Assessment of the impacts of pit latrines on groundwater quality in rural areas: a case study from Marondera district, Zimbabwe. Phys Chem Earth 31(15–16):779–788

Farooqi A, Masuda H, Firdous N (2007) Toxic fluoride and arsenic contaminated groundwater in the Lahore and Kasur districts, Punjab, Pakistan and possible contaminant sources. Environ Pollut 145(3):839–849

Franceys R, Pickford J, Reed R (1992) A guide to the development of on-site sanitation. WHO, Geneva

Gerba CP, Bitton G (1984) Microbial pollutants: their survival and transport pattern to groundwater. In: Bitton G, Gerba CP (eds) Groundwater pollution microbiology. Wiley, New York, pp 65–88

Gerba CP, Wallis C, Melnick JL (1975) Fate of waste water bacteria and viruses in soil. J Irrig Drain Div Am Soc Civ Eng 101(IR3):157–175

Girard P, Hillaire-Marcel C (1997) Determining the source of nitrate pollution in the Niger discontinuous aquifers using the natural 15N/14N ratios. J Hydrol 199:239–251

Graham JP, Polizzotto ML (2013) Pit latrines and their impacts on groundwater quality: a systematic review. Environ Health Perspect 121:521–530. https://doi.org/10.1289/ehp.1206028

Hagedon C (1984) Microbiological aspects of groundwater pollution due to septic tanks. In: Britton B, Gerba CP (eds) Groundwater pollution microbiology. Wiley, New York, pp 181–196

Howard G, Teuton J, Luyima P, Odongo R (2002) Water usage patterns in low-income urban communities in Uganda: implications for water supply surveillance. Int J Environ Health Res 12(1):63–73

Jacks G, Sefe F, Carling M, Hammar M, Letsamao P (1999) Tentative nitrogen budget for pit latrines–eastern Botswana. Environ Geol 38(3):199–203

Jha A, Shah K, Verma RJ (2012) Effects of sodium fluoride on DNA, RNA and protein contents in liver of mice and its amelioration by Camellia sinensis. Acta Pol Pharm 69(3):551–555

Kligler IJ (1921) Investigation on soil pollution and relation of the various privies to the spread of intestinal infections, Monograph No. 50. The Rockefeller Institute of Medical Research, New York

Kumar MD (2017) Market analysis: desalinated water for irrigation and domestic water use in India, prepared for securing water for food: a grand challenge for development in the Center for Development Innovation, US Global Development Lab. DAI Professional Management Services, US

Kumar MD, Shah T (2004) Groundwater pollution and contamination in India: emerging challenges. Hindu Survey of Environment, Kasturi and Sons

Kumar MD, Shah T (2006) Groundwater pollution and contamination in India: the emerging challenge. IWMI-TATA Water Policy Research Program Paper 2006/1, Gujarat, India

Kumar MD, Tortajada C (2020) Assessing wastewater management in India. Springer Briefs in Science and Technology, Singapore

Lawrence AR, Macdonald DMJ, Howard AG, Barret MH, Pedley S, Ahmed KM, Nalubega M (2001) Guidelines for assessing the risk of groundwater from on-site sanitation. Commissioned report (CR/01/142) of British Geological Survey, UK

Lewis J, Foster S, Drasar BS (1982) The risk of groundwater pollution by on-site sanitation in developing countries. International Reference Centre for Waste Disposal (IRCWD), Duebendorf, Switzerland

Li C, Gao X, Wang Y (2015) Hydrogeochemistry of high-fluoride groundwater at Yuncheng Basin, northern China. Sci Total Environ 508:155–165

Mafa B (2003) Environmental hydrogeology of Francistown: effects of mining and urban expansion on groundwater quality. Department of Geological Survey and Federal Institute for Geosciences and Natural Resources, Lobatse, Botswana

Nichols DS, Prettyman D, Gross M (1983) Movement of bacteria and nutrients from pit latrines in the boundary waters canoe area wilderness. Water Air Soil Pollut 20(2):171–180

Pujari PR, Padmakar C, Labhasetwar PK, Mahore P, Ganguly AK (2012) Assessment of the impact of on-site sanitation systems on groundwater pollution in two diverse geological settings: a case study from India. Environ Monit Assess 184(1):251–263

Sheth FJ, Multani AS, Chinoy NJ (1994) Sister chromatid exchanges: a study in fluorotic individuals of North Gujarat. Fluoride 27(4):215–219

Still DA, Nash, SR (2002) Groundwater contamination due to pit latrines located in a sandy aquifer: a case study from Maputaland. In: Water Institute of Southern Africa Biennial Conference, Durban, South Africa, May 2002

Tandia AA, Diop ES, Gaye CB (1999) Nitrate groundwater pollution in suburban areas: example of groundwater from Yeumbeul, Senegal. J Afr Earth Sci 29(4):809–822

Verheyen J, Timmen-Wego M, Laudien R, Boussaad I, Sen S, Koc A (2009) Detection of adenoviruses and rotaviruses in drinking water sources used in rural areas of Benin, West Africa. Appl Environ Microbiol 75(9):2798–2801

Vinger B, Hlophe M, Selvaratnam M (2012) Relationship between nitrogenous pollution of borehole waters and distances separating them from pit latrines and fertilized fields. Life Sci J 9(1):402–407

WHO (World Health Organization) (2006) Protecting groundwater for health: managing the quality of drinking-water sources. WHO, Geneva. http://www.who.int/water_sanitation_health/publications/protecting_groundwater/en/. Accessed 10 Aug 2021

Zingoni E, Love D, Magadza C, Moyce W, Musiwa K (2005) Effects of a semi-formal urban settlement on groundwater quality Epworth (Zimbabwe): case study and groundwater quality zoning. Phys Chem Earth 30(11–16):680–688

Chapter 8
Improving Institutional Responses to Groundwater Pollution: Use of a Drinking Water Quality Surveillance Index

8.1 Introduction

In view of the growing threat of groundwater pollution from microbial and biochemical contaminants posed by the millions of latrines in the rural areas of the country, and the complex factors governing the susceptibility or vulnerability of groundwater systems to pollution from on-site sanitation, a scientific and comprehensive approach to water quality surveillance is required to protect the hundreds of thousands of groundwater-based drinking water sources in India, against an approach that follows a uniform surveillance protocol for all regions based on simplistic considerations.

In this chapter, we will introduce the concept of drinking water supply surveillance, explain why a framework for drinking water quality surveillance is required for India, describe the development of a composite surveillance index, and compute the values of the index for the whole of Maharashtra state at the block level. The purpose of a framework is to help us identify regions and pockets that have the highest public health risks associated with pollution of drinking water source and where frequent water quality surveillance is required for protecting the sources. The basic idea is to economise on the efforts at water quality monitoring in rural areas to ensure the safety of drinking water supplies.

In the next section, we will discuss the concept of drinking water supply surveillance. This will be followed by a discussion on the need for a drinking water surveillance index in the third section. The fourth section discusses the issues related to water quality monitoring in the state of Maharashtra, especially those pertaining to water quality monitoring of the resource (i.e. groundwater), monitoring of drinking water sources and sanitary surveillance. This is followed by a discussion on the development of a framework for surveillance of drinking water quality and its computation for the state of Maharashtra in Sects. 8.5 and 8.6, respectively.

M. Dinesh Kumar et al., *Drinking Water Security in Rural India*, Water Resources Development and Management, https://doi.org/10.1007/978-981-16-9198-0_8

8.2 Drinking Water Supply Surveillance

Water supply surveillance is defined as: 'the continuous and vigilant public health assessment and oversight of the safety and acceptability of water supplies' (WHO 1976, 1993, 2004). Many millions of people, in particular throughout the developing world, use unreliable water supplies of poor quality, which are costly and are distant from their homes (WHO and UNICEF 2000). Water supply surveillance generates data on the safety and adequacy of drinking water supply in order to contribute to the protection of human health. Most current models of water supply surveillance come from developed countries and have significant shortcomings if directly applied elsewhere. There are differences not only in socio-economic conditions but also in the nature of water supply services, which often comprise a complex mixture of formal and informal services for both the 'served' and 'un-served' (Howard and Bartram 2005).

Some sections of society in the developing world enjoy water supply and other services of a quality comparable to those in developed countries, frequently at lower costs (HDR 2006; Howard and Bartram 2005). However, many households do not have access to tap connections at home. As a result, there is a widespread use of a wide variety of communal water sources. These include public taps, water sold by households with a connection and purchases from vendors (Whittington et al. 1991; Cairncross and Kinnear 1992; Howard 2001; Tatietse and Rodriguez 2001). They also include a variety of small point water supplies such as bore wells with hand pumps, protected springs and dug wells (Gelinas et al. 1996; Rahman et al. 1997; Howard and Luyima 1999). In India, communities depend extensively on private bore wells even when individual tap connections for treated water are provided by the utilities.

The data generated through well-designed and implemented surveillance programmes can be used to provide public health input into water supply improvements. The key to designing such a programme is information about the adequacy of water supplies and the health risks faced by populations at national or sub-national levels to identify areas that are vulnerable. But this is scarce in many countries (Howard and Bartram 2005). Scarce is also the information about the status of environmental sanitation conditions. This is despite significant advocacy of 'people-centred' and 'demand-responsive' approaches in recent years.

8.3 Why a Drinking Water Quality Surveillance Index?

Few published studies that address the development of water supply surveillance programmes in developing countries exist. According to a review, while most countries have some form of guidelines on water quality, these are not routinely enforced (Steynberg 2002). It suggested that often the health sector performs more monitoring than the water supply sector, but provided no evidence that systematic monitoring of water supply extended beyond utility piped systems. A recent assessment of drinking water supply surveillance by the WHO in South-East Asia Region noted that none of the countries had a comprehensive national programme of surveillance (Howard and Pond 2002). Though surveillance of piped water supplies was carried out, alternative sources and household water in urban areas were not typically included.

There are very few reported examples of surveillance programmes where there is a mix of water source type and service level, or which have addressed the targeting of vulnerable populations. Some projects tried to focus on alternative sources and household water, but were typically focused on single communities or were time-limited assessments of water (Howard 1997; Karte 2001). Poverty or vulnerable populations had not been a significant factor in the surveillance programme design.

In India, certain uniform protocols are followed by the Ministry of Drinking Water and Sanitation for water quality monitoring. These protocols concern the number of water quality testing laboratories in different administrative areas, the water quality testing facilities that should be available in these labs—the human resources, instruments, and the number of water quality parameters that these labs should test periodically, and the sampling procedures. The water quality testing capabilities of the laboratory are not decided by the water quality challenges a region or area faces but purely based on the administrative status of the area—i.e., whether it is a state-level lab, a district-level lab, or a sub-district-level lab. The WQM protocol prescribes that the state-level water quality testing laboratories of the government should have facilities to test a total of 78 water quality parameters, the district-level labs must have facilities to test 34 parameters and sub-district-level labs must have facilities to test 19 parameters.

There are about 50 lac reported public drinking water sources in the country (MoDWS 2013). Considering many unreported and/or private sources, the total number of drinking water sources in rural India may exceed 60 lacs. If these sources are to be tested twice a year (for bacteriological analysis) and once a year (pre-monsoon) for chemical analysis, 120 lac water samples will have to be tested in the country in a year. As reported by states, about 1,869 district and sub-district water-testing laboratories (including labs other than PHED labs) exist in the country, though many of them are still not fully functional (MoDWS 2013).

According to the Ministry of Drinking Water and Sanitation (MoDWS), if all such laboratories are made fully functional and each laboratory has a capacity to analyse 3,000 samples a year, the number of sources that could be tested in a year would be $3{,}000 \times 1{,}869 = 56$ lac, i.e. about 50% of the total samples required to be tested. Therefore, under the National Rural Drinking Water Programme (NRDWP), provision for setting up new sub-district-level laboratories has been made to strengthen the institutional capability. Further, the decentralized Water Quality Monitoring & Surveillance Programme started in the year 2005–06 envisages indicative testing of all drinking water sources (both public and private) using simple field test kits and only positively tested samples to be referred to district and sub-district water testing laboratories for confirmation. Sanitary inspection is also a part of this programme. The state also has public health laboratories situated at state, regional, and district levels. The Ministry has used the total number of water supply sources as the basis for assessing the institutional capacity in WQM in terms of the number of laboratories that need to be set up.

As a result, the Water Quality Monitoring Protocol developed by the Ministry of Drinking Water and Sanitation deals with the total number of laboratories that need to be set up at different levels (like state, district, and sub-district) and the number of WQ parameters each type of laboratory should test. As per the protocol, a total of 19 parameters will be tested in sub-district labs, 34 parameters in district labs, and 78 parameters in state-level labs (see Annexure IV, MODWS 2013, pp. 41–43). It describes specific requirements for monitoring drinking water quality with a view to ensure the provision of safe drinking water to the consumers. In addition, it also includes requirements for setting up laboratories at the state, district, and sub-district levels and their quality control for regular testing and surveillance of drinking water sources. In other words, it describes various elements of laboratory management practices to ensure that the data generated is comparable and scientifically correct and in a form that can then be used to result in interventions to improve water quality.

The inbuilt assumption in the Ministry's approach seems to be that Water Quality Management challenges are uniform. While it specifies the WQ parameters to be analysed by each type of laboratory, it does not make any distinction amongst the water sources being monitored in terms of the likely water contamination and pollution challenges each one pose and does not assign the types of water quality parameters to be analysed using the samples collected from individual sources. However, the reality is that there are certain regions/areas in every state which face inherent contamination problems, especially with respect to groundwater, and there are regions where both surface water and groundwater resources are vulnerable to pollution from a wide range of sources by virtue of their location vis-à-vis their proximity to polluting industries, urban centres, unique characteristics of geological formations that the schemes tap, geohydrology, proximity to sea, rainfall, and climate. Field monitoring of source water quality using test kits, as proposed by the Ministry of Drinking Water and Sanitation, will not be adequate to capture complex pollutants. Instead, in such regions, DW source monitoring for contaminants and pollutants will have to be more stringent and more frequent than that of others. Accordingly, the sub-district-level laboratories located in such regions should be equipped with better facilities for water quality surveillance. On the other hand, in regions where serious water contamination problems are very rare, it will not make economic sense from the point of public health to monitor a large number of water quality parameters, as every additional parameter considered for water quality analysis increases the manpower, chemicals, and equipment requirements and therefore the cost of water quality monitoring.

However, the current protocol does not recognize the need for differential treatment of areas and the water-testing laboratories located in those areas on the basis of drinking water safety threat or water quality management challenges.

8.4 Issues Related to Water Quality Monitoring in Maharashtra

There are 23,830,580 households in Maharashtra out of which 13,426,793 receive treated tap water and 2,752,519 receive untreated tap water, the rest receiving their water from other sources. 10,760,853 and 1,480,622 households receive treated and untreated tap water, respectively, within their premises, 2,151,295 and 1,032,575 households near their premises, and 514,645 and 239,322 households away from their premises. In rural Maharashtra, out of 13,016,652 households, 4,160,324 receive treated tap water while 2,380,144 receive untreated tap water and the rest from other sources (http://www.censusindia.gov.in/2011census/Hlo-series/HH06.html, accessed Oct 23, 2018). According to the NFHS 4 report, 91.5% of households received improved drinking water sources, out of 97.7% households in urban areas and 85.6% households in rural areas received improved drinking water (NFHS-4, 2015–16).

8.4.1 Water Resources Monitoring

There are four agencies monitoring water quality in Maharashtra, viz. the hydrology project of the Water Resources Department for surface water, the Maharashtra State Pollution Control Board, GSDA for groundwater and the Public Health Centres that monitor potability and bacteriological contamination of drinking water. The data generated by GSDA on groundwater quality is used to generate regional water quality trends. The data generated by MPCB on groundwater based on monitoring is used as point source data.

As we have seen, the index used by MPCB for assessing groundwater quality considers a total of 12 parameters are monitored by MPCB. They are: pH, total hardness, calcium, magnesium, chloride, total dissolved solids (TDS), fluoride, manganese, nitrate, iron, and sulphates. Hence again, parameters such as pesticide residues, heavy metals, arsenic, etc., are not monitored. The United States Environmental Protection Agency uses several criteria for defining water quality. They are: human health; aquatic life; nutrient load in water; wetland conservation; wildlife protection; priority pollutants; biological water quality; flow considerations; sediment benchmarks; and temperature. It is evident that water parameters to be monitored also depend on the purpose for which water is meant to be used (USGS 2015).

The Field Manual for Collection of Water Quality Data prepared by US Geological Survey discusses the field measurement of following water quality parameters: (1) turbidity; (2) dissolved oxygen; (3) temperature; (4) specific electrical conductance; (5) pH; (6) reduction–oxidation potential; and, (7) alkalinity and acid-neutralizing capacity. It also discusses the testing procedure for many parameters in the laboratory. They are: five-day BOD; faecal virus and bacteria; algal biomass; protozoan pathogens; and cyno-bacteria in lakes and reservoirs. The other tests discussed in the manual are for pharmaceutical compounds; antibiotic compounds; organic compounds; low-level mercury; and arsenic (Please see the National Field Manual for Collection of Water Quality Data by USGS (2015).

It is evident that the nature of use of ground water in India keeps changing from multiple uses (like in Kerala) to just water for irrigation in large parts of western India, where the resource faces high incidence of mineral contamination. A similar pattern is witnessed in Maharashtra, which displays high spatial heterogeneity in geological, hydrological, climatic, and socio-economic conditions (GSDA, IRAP and UNICEF 2013).

On the other hand, in many situations, there is only one source available for meeting different water needs, including drinking. In such situations, it is important to make sure that the available water meets the highest quality standards (to address the concerns of human health), or there is no further deterioration of the quality of water. However, the parameters chosen for water quality monitoring by Maharashtra Pollution Control Board remain more or less the same, and the difference in monitored WQ parameters is noticed only between surface water and groundwater (Source: based on MPCB 2017). The water quality parameters analysed under the hydrology project for surface water are the same for all the 127 stations, except for the two reservoirs for which two additional parameters are monitored.

Further comparison between the surface water quality parameters analysed by MPCB and those by the hydrology project show that the number of parameters analysed by the latter is much more than that by the former (32 against four). The frequency of sampling water is also not the same—once in a month in the case of MPCB stations against once in a fortnight for most locations (77 out of 127 locations) in the case of hydrology project. This inconsistency makes use of integrated data set for further analysis of water quality difficult. In the case of groundwater, though the number of parameters analysed is more or less the same, 50% of the parameters are different between GSDA and MPCB. But the water quality testing does not consider analysis for heavy metal content in water samples in either of the cases.

8.4.2 Monitoring of Drinking Water Sources

Given the fact that many of the drinking water sources are individual village schemes and point sources (handpumps) often serving a small group of population, a large number of drinking water sources had to be monitored in Maharashtra every year to ensure safe water supply. As we have seen, a total of around 187,000 sources are currently being monitored.

Like in other parts of India, water quality monitoring for drinking water safety in the state does not follow any such 'risk-based approach'. This is evident from the facilities proposed for water quality testing laboratories and the parameters tested.

The facilities prescribed for the labs depend on whether there are state-level or district-level or sub-district-level labs, and not based on the water quality issues facing the locality. The best facilities are prescribed for state-level labs, with equipment for testing 78 parameters, followed by district-level labs for testing 34 parameters, though this norm is not followed by the state of Maharashtra. Similarly, the norms for staffing of labs at the district level are not determined by the nature and magnitude of the water quality problems in the districts.

Further, irrespective of the location, water samples are being collected periodically (twice a year) and tested for certain basic parameters. But the test results of water samples collected from consumer taps or supply points of drinking water sources are not compared with those of source water to understand: the vulnerability of DW sources to contamination, or as to what extent source water quality monitoring can be used to predict the quality of the water from various drinking water supply points. Secondly, the results are not compared against the regional environmental conditions to analyse the latter's effect on drinking water quality.

Such an approach followed in water quality monitoring increases the cost of monitoring per unit population. Again, the sources keep changing as some of the sources become dysfunctional due to mechanical problems or (groundwater) resource depletion, and new sources are created in different locations. This further complicates the matter as long-term monitoring of the sources to study temporal changes becomes infeasible due to various extraneous factors.

Over and above the issue of cost, there appears to be a major shortage of human resources for water quality monitoring at the level of laboratories for analysis of water samples. There are 172 water quality testing laboratories in the state—from one at the state level to one each at the district level to block-level laboratories in most of the blocks. Out of these, six labs are at the division level and 44 at the district level. However, state-level lab is not functional. Another 11 labs are under construction. A large proportion of the technical staff in these laboratories (chemists and microbiologists) are in major cities such as Thane, Pune, and Nagpur.

Table 8.1 shows the number of district-level laboratories and their staffing pattern. From the table, it is clear that many of the laboratories are understaffed as the minimum number of technical staff required in these labs is eight. In five district laboratories, the position of microbiologist is vacant. In 14 of the labs, there are no lab assistants. Almost one-third of the staff (125 out of 344) are non-technical staff. Further, a quick review of the regional (ground) water quality problems and the distribution of staff shows that the staffing of labs at the district level is not determined by the nature and magnitude of the water quality problems in the districts. Hence, the ability of many of the laboratories at the district level and below to carry out the job of water quality testing and do proper interpretation is questionable. There are a number of labs that do not have microbiologists.

Table 8.1 Staffing composition of district-level water quality testing laboratories in Maharashtra

District name	No. of Labs available	No. of chemists	No. of bacteriologists	No. of Lab assistants	Others	Total staff
Thane	3	10	5	4	12	31
Raigad	1	2	2	1	3	8
Ratnagiri	1	2	1	2	1	6
Sindhudurg	1	2	2	0	4	8
Nashik	2	5	2	2	8	17
Dhule	1	2	1	1	3	7
Jalgaon	2	4	2	2	2	10
Ahmednagar	1	3	1	3	3	10
Pune	2	12	10	5	11	38
Satara	1	2	2	0	1	5
Sangli	1	3	2	2	4	11
Solapur	1	3	2	2	3	10
Kolhapur	1	4	1	0	5	10
Aurangabad	2	3	0	0	4	7
Jalna	1	2	2	0	4	8
Parbhani	1	3	1	2	5	11
Beed	1	2	2	0	4	8
Nanded	1	1	0	0	1	2
Osmanabad	1	1	0	0	1	2
Latur	1	2	2	2	2	8
Buldana	1	2	3	2	3	10
Akola	2	3	1	0	6	10
Amravati	2	4	2	1	5	12
Yavatmal	2	2	1	1	5	9
Wardha	1	2	2	1	1	6
Nagpur	2	10	5	5	3	23
Bhandara	1	3	2	2	2	9
Chandrapur	2	5	0	2	5	12
Gadchiroli	1	2	1	0	5	8
Nandurbar	1	3	1	2	1	7
Washim	1	2	4	0	2	8
Gondia	1	2	2	0	3	7
Hingoli	1	2	1	0	3	6
Palghar	0	0	0	0	0	0
	44	110	65	44	125	344

Source Figures based on August 2018 data, accessed from the NRDWP website (https://ejalshakti.gov.in/IMISReports/NRDWP_MIS_NationalRuralDrinkingWaterProgramme.html)

8.4.3 *Sanitary Surveillance*

Since 2013, Maharashtra is periodically undertaking the assessment of public health risks posed by contaminated public drinking water sources in rural areas using sanitary surveys as per the guidelines issued by NRDWP in 2012. The assessment has so far covered 80% of the total drinking water supply sources in the state and has shown that the total number of 'high-risk' and 'medium-risk' sources has come down. The proportion of GPs receiving a green card has actually gone up from 67.6% in April 2013 to 78% in October 2017. The proportion of GPs receiving yellow cards has also come down from 29.6% in April 2013 to 18.5% in October 2017. There is a drastic reduction in the number of GPs receiving a red card, showing marked improvement in the drinking water safety situation. In terms of the number of sources, the number of low-risk sources had increased from 70.6 to 80% during the same period, a marked improvement of more than 9%.

The methodology used for assessing public health risks associated with contaminated DW sources based on sanitary surveys is suggested by the MoDWS, under the Uniform Protocol. More or less the same methodology is followed in Maharashtra by WSSO (MoDWS 2013). Nevertheless, additionally, in Maharashtra, water quality samples from the sources are collected and analysed for bacteriological contamination, an important aspect that is missed in the NRDWP protocol. This issue notwithstanding, a detailed review of this methodology raises the following issues. Though the methodology factors in the unique characteristics of different water sources—reservoir, bore well, open well, HP, spring, etc., and several sets of criteria (diagnostic information) are considered for assessing risk, they are narrow technical in nature and several of them are arbitrary (distance of the DW source from latrines), and do not have much bearing on the risk DW supply sources can pose when considered in isolation. Some key parameters, which define environmental conditions around the water source, which have a significant bearing on the chances of contamination of the water source from contaminants and pollutants, are not considered. They are: depth to groundwater table, quality of underground water in the surrounding area, discharge of pollutants in the channel (for groundwater-based sources), and rainfall, humidity, topography, soil conditions, and climate.

In the absence of a comprehensive analytical framework that incorporates the whole range of physical, socio-economic and environmental factors influencing the public health risks associated with contamination of drinking water sources from insanitary conditions, the sanitary surveys will turn out to be prohibitively expensive, if every GP and every drinking water source is to be covered periodically for ensuring drinking water safety. Moreover, the survey results may be misleading, as they may not reveal the real danger of future occurrence of hazardous events in terms of contamination of drinking water. Therefore, a simple index, which can provide indications on the 'risk-prone areas' that require water quality and sanitary surveillance, needs to be developed for the state.

8.5 Development of a Drinking Water Quality Surveillance Index for Assessing the Risk of Drinking Water Sources

The water quality surveillance index uses several of the concepts used in the development of the WATSAN vulnerability index developed for urban areas by Kumar (2014), which is about assessing the vulnerability of communities to problems associated with the lack of adequate quantities of water of sufficient quality and reliability for domestic and productive needs. The theoretical discussions providing the rationale for using several of the parameters considered for assessing 'threat', 'exposure', and 'vulnerability' are provided in Kumar (2014).

The development of the present index considers the factors that determine:

1. The 'threat' to water sources, or the chances of deterioration (through contamination or pollution) of water source (such as groundwater or surface water), which is captured by a water quality index for the source water
2. The degree of exposure of the drinking water sources to contamination or pollution, which is determined by four key factors, viz. availability of water in terms of quantity and quality; conditions of water supply infrastructure; access to water and sanitation; and climate, flood proneness and population density, and,
3. The vulnerability of the communities, which is influenced by two factors, viz. overall public health status, and institutions and management.

The total risk is computed by multiplying Threat (T) X Exposure (E) X Vulnerability (V).

The maximum value of each sub-index would be 1.0. The value of each sub-index is computed by adding up the values of various factors that influence it and then normalizing to obtain a maximum value of 1.0. For computing the value of the sub-index for 'threat', the 'water quality index' (of the source water), the only parameter which influences it is used. For computing the value of the sub-index for 'exposure', values of variables, A, B, C, and G are added up and normalized. For computing the value of 'vulnerability' sub-index, values of variables, D and F are added up. Lower values of the index meaning higher vulnerability. The factors considered for computing the sub-indices also will have equal weightage (measured on a scale of 0 to 1.0) and the sum of their values will have to be normalized so as to obtain a maximum value of 1.0. Here it is important to reckon with the fact that the values being assigned to the indices and sub-indices used for computing the water quality surveillance index, many of the variables used for computing various indices (say, for instance, the climate sub-index) are 'discrete distributions', and are NOT 'real value functions' of real variables, and this is used because what we have developed is a statistical model for computing the surveillance.

The index is computed separately for groundwater-based sources and surface water-based sources, as two of the parameters, viz. resource vulnerability to pollution/contamination, and water quality index values, are specific to the type of resources under consideration (i.e., whether surface water or groundwater). Which index value we shall use for assessing the water quality surveillance needs will depend on the ground situation. If there are only surface water schemes in an area, then the index corresponding to surface water shall be used. If there are only groundwater-based schemes in an area, then the index corresponding to groundwater shall be used. In the case of the combined use of groundwater and surface water, the weighted average on the basis of the population served by each type of source shall be used.

Though similar indices are not in use anywhere, in the UK, a Compliance Risk Index (CRI) is used by the Water Supply Inspectorate to assess the risk associated with non-compliance of the European Commission's regulations on drinking water supply by the water supply companies. The Compliance Risk Index is a measure designed to show the risk arising from treated water compliance failures, and it is in line with a risk-based approach to the regulation of water supplies. It tries to quantify the significance of each parameter in deciding the overall risk associated with non-compliance of water quality standards, the proportion of consumers potentially affected and an assessment of the company response. The CRI helps the water supply companies to request (to the Inspectorate) for making adjustments in their sampling programme (DWI 2018).

In the USA, the US Environmental Protection Agency uses the concept of risk assessment of 'source water' for designing the source water monitoring system. This risk assessment is based on likelihood, vulnerability, and consequences. Likelihood is the probability that the source water will get contaminated. Its value may be based on previous contamination incidents caused by the source water threat or on projections and models. Vulnerability is the probability that a utility or its customers would be impacted by a source water threat. The vulnerability value is generally based on the ability of the utility to effectively respond to an SW threat, preventing or mitigating consequences to utility infrastructure, operations, and customers. Consequences are the adverse effects of an incident experienced by a utility or its customers (e.g. illness). Where possible, consequences are expressed in terms of monetary damage, providing a standard measure of consequence across all threats (USEPA 2016).

The various parameters considered for developing the water quality surveillance index, the rationale for considering them, the method of putting scores for each of the parameters required for computing the value of the index, and the sources of obtaining the data corresponding to these parameters are detailed in Table 8.2.

8.6 Drinking Water Quality Surveillance Index for Maharashtra

Figure 8.1 depicts the computed values of WQSI for different blocks. *Shegaon* block in Buldhana district is at the highest risk (DWQSI of 0.090) and *Panhala* in Kolhapur district is at the lowest risk (DWQSI of 0.582). Overall, 77 blocks are in the high-risk category (DWQSI ranging from 0.064 to less than 0.216) which constitutes 22% of all the blocks of the state (shown in red in the map); 273 blocks in the moderate risk category (values of DWQSI value ranging from 0.216 to less than

Table 8.2 Various parameters considered for developing the water quality surveillance index

No.	Parameters	Rationale	Scoring Methods	Data Source
Threat				
A	Water Quality Index			
1.0	Water Quality Index (The type of resource to be considered, i.e. whether groundwater or surface water, for computation of this index, should be based on the types of drinking water sources, which require surveillance. If the source is a reservoir, then only the quality status of surface water should be considered. But if the source is a well or handpump or tube well, then the overall quality of SW and groundwater shall be considered)	Sources with the degree of water pollution and differentiates the purity and impurity of the water resources	Water Quality Index (Potable = 1.0, non- potable = 0.0). This is based on the WQI adopted and estimated by the Central Pollution Control Board	The maximum score of sub index E will be 1.0
Exposure				
B	Availability of Water Resources in Terms of Quantity and Quality			

(continued)

Table 8.2 (continued)

No.	Parameters	Rationale	Scoring Methods		Data Source
1.1	Surface water and groundwater availability in the area	When plenty of freshwater is available, the chances of communities resorting to consuming water from contaminated sources is less A renewable freshwater availability of 1700 m^3 per capita per annum is considered adequate for a region or town, estimated at the level of the river basin in which it is falling	The value of the index is computed by taking the amount of renewable water as a fraction of the desirable level of 1,700 m^3	The maximum score under sub-index A will be 3.0. It has to be reduced to 1.0 for normalizing	WRD, Govt of Maharashtra
1.2	Variability in resource condition	Higher the variability, greater will be the vulnerability	The index is computed as an inverse function of the coefficient of variation in the rainfall variability in that basin/sub-basin $(1 - x/100)$; where x is the coefficient of variation in rainfall		IMD/ maharani.gov.in
1.3	Seasonal variation of groundwater	With the presence of alluvial aquifers, the vulnerability lowers, and in the presence of hard rock aquifers, vulnerability will be high. Also, in regions with sedimentary and alluvial deposits, vulnerability will be medium	For alluvial areas, the value of this index is considered as 1. For hard rocks, the value is considered as 0.3. For sedimentary and alluvial deposits, the value is considered as 0.65		GSDA/ CGWB

(continued)

Table 8.2 (continued)

No.	Parameters	Rationale	Scoring Methods		Data Source
2.0	Vulnerability of the resource to pollution or contamination Here, in the case of sources based on groundwater, both surface and groundwater resources shall be considered. But in the case of surface water-based sources, only the vulnerability of SW resources shall be considered	Shallow groundwater; river/stream/reservoirs in the vicinity of industries are highly vulnerable	Shallow groundwater, river/stream/reservoirs in the vicinity of industries are highly vulnerable with a value of the sub-index equal to 0.0; distant reservoir in the remote virgin catchments and groundwater from deep confined aquifers has a pollution vulnerability index of 1.0; shallow groundwater in rural areas (with no industries) to have medium vulnerability with a value of 0.50		CPCB
3.0	Pollution from Industrial and Domestic Sources	In the presence of different categories of industries (pulp and paper, distillery, power, tannery, dyeing industry, etc.), pollution will be high. Also, a higher population causes wastewater discharge to be high	Presence of highly polluting industries near rivers/streams and population more than 100,000 = 0.0, if industries not present near rivers/lakes and population less than 100,000 = 1.0		CPCB/HP
C	Accessibility				
1.0	Percentage of households with piped water supply having access to treated (tap) water	Reduces chances of contamination of water during collection and storage	Index will be computed based on % households having access to piped water supply estimated as a fraction of total households	The maximum score under sub-index B will be 3.0 and has to be reduced to 1.0 for normalizing	Census

(continued)

Table 8.2 (continued)

No.	Parameters	Rationale	Scoring Methods		Data Source
2.0	Access to sanitation as a % of total households	Reduces chances of vector borne diseases through food contamination, etc.	% of households with piped water supply having access to treated (tap) water		Census
3.0	Percentage of households depending on public taps/stand post/ and other sources (tanks/lakes, etc.),	Chances of contamination of water will be higher for the open spaces (tanks/lakes) also during collection and storage	Percentage of households depending on public taps/stand post, value of which is estimated as an inverse function		Census
D	Infrastructure characteristics				
1.0	Percentage of households covered by the water distribution system	Reduces chances of contamination of water during collection and storage	Percentage of households covered by the water distribution system as fraction	The maximum score under sub-index C will be 4.0 and has to be reduced to 1.0	WSSO/PHED/Census
2.0	Percentage of households covered by sanitation system	Reduces chances of vector-borne diseases through food contamination, etc.	Percentage of households covered by sewerage system as fraction		WSSO/PHED/Census
3.0	Condition of water supply infrastructure	Old water supply systems are more susceptible to disruption	very good = 1.0; good = 0.8; average = 0.6; poor = 0.4; very poor = 0.20		WSSO/ PHED
4.0	Condition of sanitation systems	Old sanitation systems are more susceptible to disruption	very good = 1.0; good = 0.8; average = 0.6; poor = 0.4; very poor = 0.20		WSSO/ PHED
E	Climate, flood proneness, and population density				

(continued)

Table 8.2 (continued)

No.	Parameters	Rationale	Scoring Methods		Data Source
1.0	Climate (whether semi-arid/arid/hyper-arid or sub-humid/humid)	The vulnerability to poor environmental sanitation is a function of climate. It increases from hot and arid to hot and semi-arid to hot and sub-humid to hot and humid to cold and humid	The value ranges from '0.0' for cold and humid to '1.0' for hot and arid with increments of "0.20"	The maximum score under sub-index G will be 3.0 and has to be reduced to 1.0	DPAP/IMD
2.0	Flood proneness (whether flood-prone or not)	Vulnerability increases with increase in flood proneness	The value can be '0.0' for flood-prone area and '1' for the rest		NIDM
3.0	Population Density		Relative index, computed by taking inverse of the value in relation to the best block		Census
Vulnerability					
F	Public health outcomes				
1.0	Under-five mortality rate (IMR) (out of 1,000 people)	Undernourishment in general and malnourishment, especially among children, make community more vulnerable	A relative index, value of which is computed by taking the inverse of the value in relation to best block $\{IMR_{max}-IMR\}/(IMR_{max}-IMR_{min})$	The maximum score under sub-index D will be 2.0 and has to be reduced to 1.0	Health Department, Govt of Maharashtra
2.0	Percentage of households reporting illness due to WRD (diarrhoea, skin rashes, worm infestation)	In sources with lower biological contamination, WRD will be less	A relative index, computed by taking inverse of the value in relation to the best block $\{WRD_{max}-WRD/[WRD_{max}-WRD_{min}\}$		Health Department, Govt of Maharashtra
G	Institutions and management				

(continued)

Table 8.2 (continued)

No.	Parameters	Rationale	Scoring Methods		Data Source
1.0	Staff: Number of staff per 1,000 connections	Vulnerability reduces with increase in number of staff members available for managing a fixed number of connections	Relative index, value of which is computed on the basis of the highest and lowest numbers found across utilities	The maximum score under sub-index F will be 4.0 and has to be reduced to 1.0	
2.0	Frequency of Complaints	Vulnerability increases with number of complaints about water supply	Vulnerability increases with number of complaints about water supply		
3.0	Time Taken to handle complaints	Vulnerability increases with delay in handling complaints	Relative index, computed by taking time taken by the best and worst performing utilities		
4.0	Performance Improvement Measures	(a) Leak detection; (b) Leakage reduction; (c) Computerization of customer care; (d) online payment and complaint registration; (e) use of GIS in planning and data management; (f) performance rewarding; (g) autonomy in hiring and firing staff	Presence of these management measures reduces the vulnerability of the city to poor water and sanitation. Presence of each one of them in the management would earn the utility a score of 1/7. In the absence of it, the score would be 0.0		

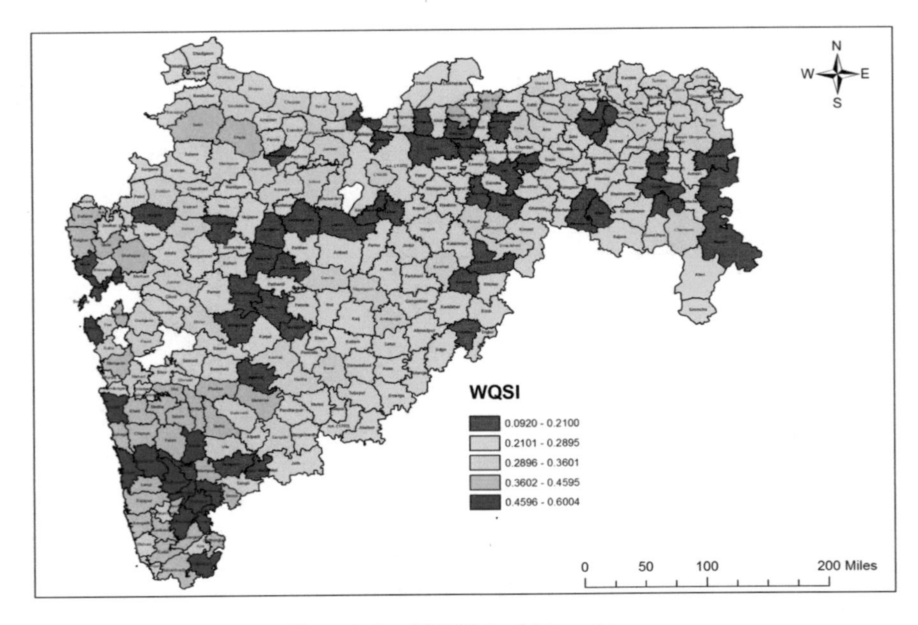

Fig. 8.1 Water quality surveillance index (WQSI) for Maharashtra

0.512), and only 4 blocks are in the low-risk category (DWSQI value greater than 0.512) (please refer to Fig. 8.1). Sixty three out of the 73 high-risk category blocks are concentrated in Amravati, Aurangabad and Nagpur divisions, which receive lesser rainfall as compared to other divisions of the state and are underlain by hard rock aquifers with limited groundwater storage potential. The low-risk blocks are *Panhala*, *Radhanagari*, *Hatkanangle* and *Karveer*, all in the Kolhapur district of Pune division, which is better off in terms of rainfall and surface water availability.

In order to understand the reasons for the high variation in public health across blocks, it is important to also analyse the parameters that influence the three sub-indices of the DWQSI, which ultimately decide on the degree of risk. These sub-indices are: 'threat' to water sources or the chances of deterioration (through contamination or pollution) of water quality; degree of 'exposure' of the drinking water sources to contamination or pollution of the resources that are tapped; and vulnerability of the communities to health hazards associated with consuming water from contaminated sources.

For analysing the threat to drinking water sources, water quality index (WQI) was considered. In most of the rural areas of the state, drinking water sources are based on shallow groundwater. Since groundwater quantity and quality can be influenced by the anthropogenic pressure in the catchment (for instance, discharge of untreated effluent over land or in reservoir or river can pollute the underlying aquifer as well), it is important to consider the quality of both surface water and groundwater resources. Water quality index values based on data from 251 surface water monitoring sites and 3694 groundwater monitoring locations were considered. It was found that WQI is excellent for 50 of the blocks, i.e. WQI value of 63–100 for surface water and less than 50 for groundwater resources. These blocks were distributed among Amravati and Nandurbar (one block each), Satara and Thane (two blocks each), Nasik and Pune (three blocks each), Kolhapur and Palghar (five blocks each), Sindhudurg (eight blocks), Ratnagiri (nine blocks), and Raigad (11 blocks) districts. Half of these blocks are in the Konkan division of Maharashtra, where the quality of surface water and groundwater resources is much better (except for those in Mumbai City and Mumbai sub-urban district) than the blocks in other divisions of the state. Further, out of the four blocks in the overall low-risk category, two blocks (*Panhala* and *Radhanagari*) were having excellent WQI.

A significantly large number of blocks (215) were in good WQI (value of 50–63 for surface water and 50–100 for groundwater resources). Nevertheless, 89 blocks were found to have poor or very poor WQI (value in the range of 0–50 for surface water and 100–300 for groundwater), worst being the *Shegaon* block in Buldhana district, *Daryapur* block in Amravati district, and *Jafrabad* block in Jalna district. *Shegaon* block is also having the highest overall risk index score. Figure 8.2 shows the variation in pollution/contamination threat to water sources. The blocks facing a high degree of 'threat' are shown in red.

For analysing the degree of exposure of the drinking water sources to contamination or pollution, four factors were considered, viz. availability of water resources in terms of quantity and quality; accessibility to water sources; infrastructure characteristics; and climate, flood proneness, and population density. Overall, exposure was low (sub-index value greater than 0.70) for 31 blocks, spread across seven districts.

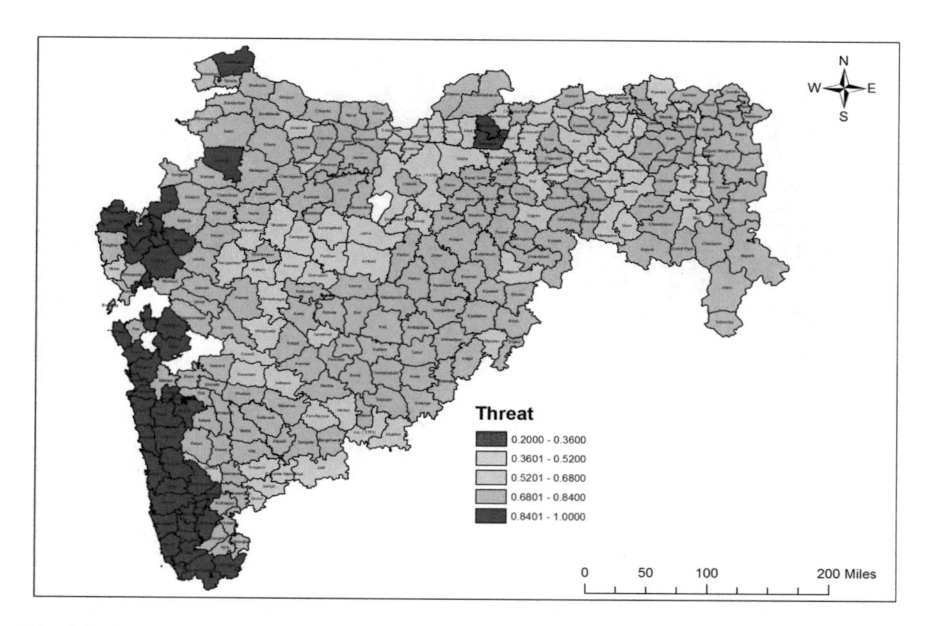

Fig. 8.2 Threat to water sources in rural areas of Maharashtra

These include Sindhudurg and Buldhana (one block each), Amravati (three blocks), Raigadh (four blocks), Satara (five blocks), Ratnagiri (six blocks), and Kolhapur (11 blocks). Out of these, 16 blocks are in the Pune division and 11 are in the Konkan division. Among them is also Shegaon (exposure sub-index value of 0.772) which faces very high overall risk. Low exposure of the two divisions can be attributed mainly to: (1) low vulnerability of water resources to pollution from the industrial and domestic wastewater in a majority of the blocks in Konkan (except for those in Mumbai City and Mumbai sub-urban district) and Pune division (except for Pune district); and (2) better household access to treated piped water supply (38% in Pune and 27.2% in Konkan division, state average is only 25.5%) and sanitation facilities (70% in Pune and 60% in Konkan division, state average is only 40%) reducing chances of contamination of water during collection and storage, and vector-borne diseases through food contamination, respectively.

Nevertheless, about 56 blocks have exposure sub-index value below 0.5. They include blocks in Akola, Aurangabad, Nandurbar, and Osmanabad (1 each); Beed, Chandrapur, and Hingoli (2 each); Buldhana and Parbhani (4 each); Gadchiroli, Nasik, and Washim (6 each); Latur (7); and Ahmednagar (12) districts. These blocks were mainly spread across Aurangabad, Amravati, and Nagpur divisions. These are the divisions with poor household access to treated piped water supply (only 19.4% of the households) and sanitation facilities (only 27.3% of the households). Please refer to Fig. 8.3 for understanding the spatial variation in the degree of exposure of the drinking water sources to contamination and pollution. The blocks facing a high

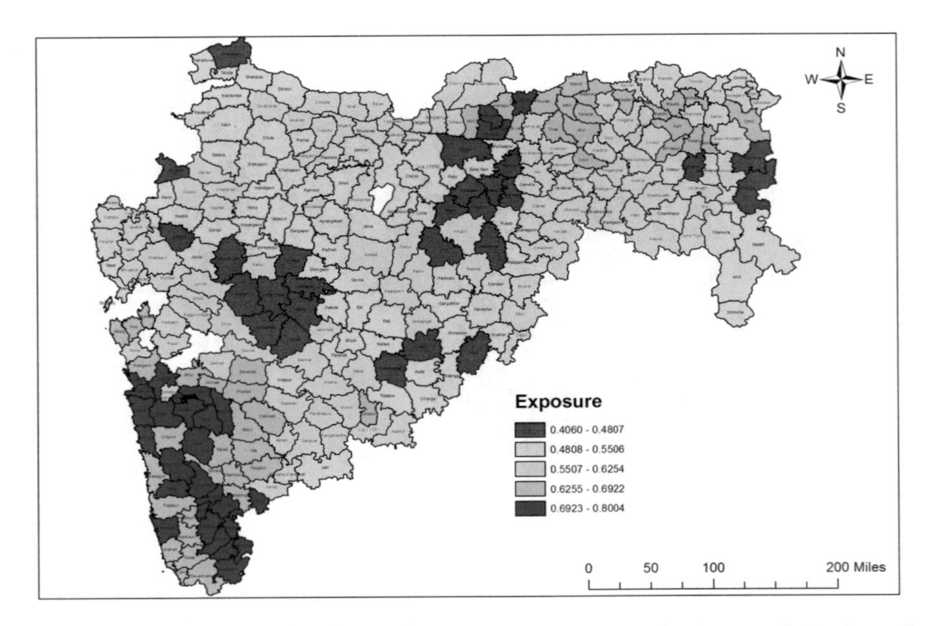

Fig. 8.3 Degree of exposure of the drinking water sources to contamination or pollution in rural areas of Maharashtra

degree of exposure are shown in red, and those experiencing moderate degrees are shown in deep yellow.

In order to determine the vulnerability of the communities to problems of exposure to contaminated water sources, two factors were considered, viz. overall public health status, and institutions and management. In 56 blocks, vulnerability (value of the sub-index greater than 0.7) was found to be low, and in 99 blocks, it was found to be high (sub-index value between 0.4 and 0.55). Additionally, the *Pune City* block was found to be highly vulnerable (sub-index score of 0.23). In 198 other blocks, the vulnerability was moderate (sub-index value ranging from 0.56 to 0.70). Figure 8.4 shows the spatial variation in the vulnerability of the communities to public health risks associated with exposure to contaminated water sources. The blocks experiencing a high degree of vulnerability are shown in red, and those experiencing a moderate degree of vulnerability are shown in deep yellow.

Among the blocks with lower vulnerability, seven each were in Ahmednagar and Kolhapur districts, and four each in Aurangabad, Buldhana, Dhule, Nashik, Satara, and Solapur districts. Most of these blocks are in Nashik and Pune divisions. *Hatkanangle* block of Kolhapur district has the lowest vulnerability (sub-index score of 0.960) mainly due to low (13.5) under-five mortality rate (state average is 14.5) and a small proportion of the population (158 per block) affected by water related diseases annually (state average is 352 per block). Even the Konkan division, which has low threat and exposure, has an under-five mortality rate of 14.2 and 355 people per block affected by water-related diseases.

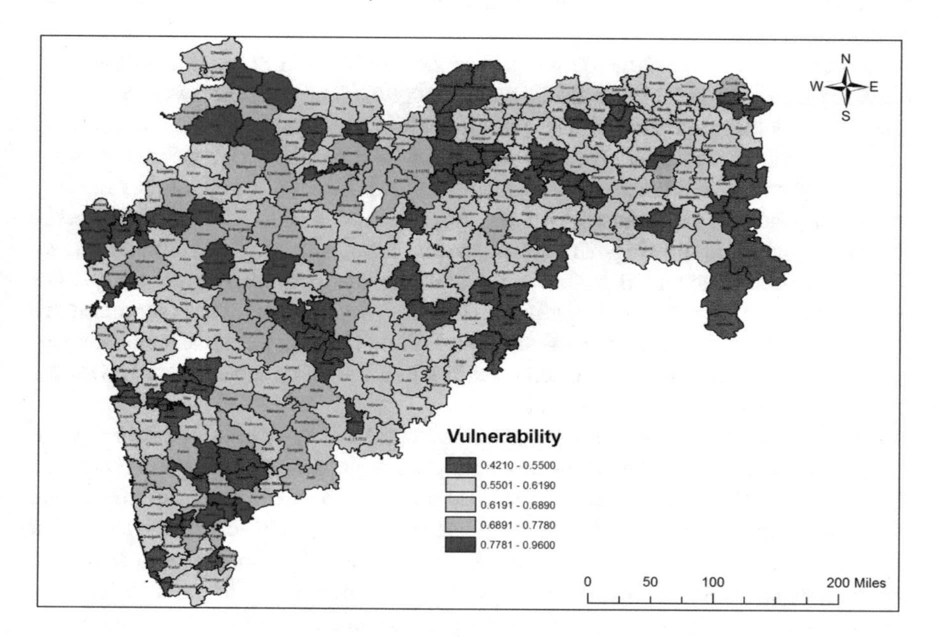

Fig. 8.4 Vulnerability of the communities to contaminated water sources

Most of the blocks with high vulnerability were spread across Aurangabad, Amravati, and Nagpur divisions. *Lonar* block in Buldhana district was found to be having the lowest sub-index score (0.421), indicating high vulnerability mainly due to a high under-five mortality rate (90) and thus making the community (including children) highly vulnerable to problems associated with consuming water from contaminated sources.

8.7 Conclusions

Through the example of Maharashtra, we have shown that the protocol for drinking water quality surveillance in India, as prescribed by the erstwhile Ministry of Drinking Water and Sanitation and currently followed by the state governments, is not only inefficient but also unscientific. The inbuilt assumption in the Ministry's approach seems to be that water quality management challenges are uniform. While it specifies the water quality parameters to be analysed by each type of laboratory, it does not make any distinction amongst the water sources being monitored in terms of the likely water contamination and pollution challenges each one pose and does not assign the types of water quality parameters to be analysed using the samples collected from individual sources. Hence, it is not based on an assessment of real public health risks posed by the contamination of drinking water sources and therefore does not prioritize areas where water quality monitoring has to be rigorous and frequent.

A water quality surveillance index was developed with particular reference to ensuring safe and potable drinking water supplies. The purpose was to identify regions/areas that pose the highest public health risks from exposure to contaminated or polluted water for drinking and that which require frequent and rigorous monitoring and surveillance of water resources and drinking water sources in order to prevent water-borne diseases, with the aim of economizing the investments made for water quality monitoring. The development of the index was based on an assessment of 'risk' that considers pollution threat to the source water, degree of exposure of the drinking water system to pollution, and the vulnerability of the community to the disruptions in the services caused by pollution. The unit of assessment of water quality surveillance index considered is a block or taluka or sub-district. Lower the value of the index, the higher the risk.

The data required for computing the index will have to be obtained from a wide variety of sources, including the water resources dept., water supply and sanitation department (MJP, WSSO, GSDA), the state pollution control board, and the public health department. The variables for deriving the index were chosen in such a way that their values could be obtained from secondary sources. The robustness of the index can be gauged from the data on the incidence of water-borne diseases in different regions/areas. Ideally, the regions showing lower values of the surveillance index would experience a higher incidence of water-borne diseases. The same index can also be used for public health surveillance to monitor the outbreak of diseases during weather-induced hazards such as floods, cyclones, and severe droughts, which cause disruptions in water supply vis-à-vis the quantity and quality of water, and taking measures for preventing contamination of water supplied from public drinking water systems, such as setting up of raw water treatment systems for the source water.

A higher value of the index suggests lower risk and vice versa. The composite DWQSI and the indices for 'threat', 'exposure', and 'vulnerability' were computed for all blocks of Maharashtra. Shegaon block in Buldhana district is at the highest risk (DWQSI of 0.090), and Panhala in Kolhapur district is at the lowest risk (DWQSI of 0.582). Overall, 77 blocks are in the high-risk category (DWQSI ranging from 0.064 to less than 0.216), which constitutes 22% of all the blocks of the state (shown in red in the map); 273 blocks in the moderate risk category (values of DWQSI value ranging from 0.216 to less than 0.512), and only 4 blocks are in the low-risk category (DWSQI value greater than 0.512). The analysis suggests that the blocks/talukas having low values of DWQSI (like Shegaon block in Buldhana) require frequent and rigorous monitoring of source water for water quality parameters.

References

Arghyam, Institute for Resource Analysis and Policy (2009) Deriving a WATSAN vulnerability index for urban areas at the household level. Institute for Resource Analysis and Policy, Hyderabad, India

Asian Development Bank (2007) 2007 benchmarking and data book of water utilities in India. Ministry of Urban Development, Government of India and the Asian Development Bank, Philippines

Cairncross S, Kinnear J (1992) Elasticity of demand for water in Khartoum, Sudan. Soc Sci Med 34(2):183–189

Drinking Water Inspectorate (2018) Drinking water 2017: summary of the chief inspector's report on drinking water in England. Drinking Water Inspectorate, London, UK

Fass SM (1993) Water and poverty: implications for water planning. Water Resour Res 29(7):1975–1981

Gelinas Y, Randall H, Robidoux L, Schmit JP (1996) Well water survey in two districts of Conakry (Republic of Guinea), and comparison with the piped city water. Water Res 30(9):2017–2026

Government of India (1999) Integrated water resources development: a plan for action, report 1. National Commission for Integrated Water Resources Development, Ministry of Water Resources, Government of India, New Delhi

GSDA (Groundwater Survey and Development Agency), Institute for Resource Analysis and Policy (IRAP) and UNICEF (2013) Multiple-use water services to reduce poverty and vulnerability to climate variability and change, A Collaborative Action Research Project in Maharashtra, India, Final Report, January 2013.

Howard G (1997) Water-quality monitoring and NGOs. Waterlines 16(1):19–22

Howard G (2001) Challenges in increasing access to safe water in urban Uganda: economic, social and technical issues. In: Craun GF, Huachman FS, Robinson DE (eds) Safety of water disinfection: balancing microbial and chemical risks. ILSI Publications, Washington, DC, pp 483–499

Howard G, Luyima PG (1999) Urban water supply surveillance in Uganda. In: Pickford J (ed) Integrated development for water supply and sanitation. 25th WEDC Conference, Addis Ababa, Ethiopia. WEDC, Loughborough, pp 290–293

Howard G, Bartram J (2005) Effective water supply surveillance in urban areas of developing countries. J Water Health 3(1):31–43

Howard G, Pond K (2002) Drinking water surveillance programmes in the south-east Asia region: updated situation assessment and recommendations for future activity. World Health Organization South-East Asia Regional Office, New Delhi, India
Human Development Report (2006) Human development report-2006. United Nations, New York
Karte D (2001) Drinking water contamination in Kolkata. In: Pickford J (ed) Water, sanitation and hygiene: challenges of the millennium. 26th WEDC conference, Dhaka, Bangladesh. WEDC, Loughborough, pp 224–226.
Kumar MD (2014) Thirsty cities: how Indian cities can meet their water needs. Oxford University Press, New Delhi
Laurence M, Sullivan C (2003) Water poverty of nations: international comparisons. Kellee University, Wallingford
Lloyd BJ, Bartram JK (1991) Surveillance solutions to microbiological problems in water quality control in developing countries. Water Sci Technol 24(2):61–75
Ministry of Drinking Water and Sanitation (MoDWS) (2013) Uniform drinking water quality monitoring protocol, Ministry of Drinking Water and Sanitation, Government of India, February 2013
National Environmental Engineering Research Institute (2005) Impact of onsite sanitation systems on ground and surface water resources. NEERI, Nagpur, India
Rahman A, Lee HK, Khan MA (1997) Domestic water contamination in rapidly growing megacities of Asia: case of Karachi, Pakistan. Environ Monit Assess 44(1):339–360
Stephens C, Akerman M, Avle S, Maia PB, Campanario P, Doe B, Tetteh D (1997) Urban equity and urban health: Using existing data to understand inequalities in health and environment in Accra, Ghana and Sao Paulo, Brazil. Environ Urban 9(1):181–202
Steynberg M (2002) Drinking water quality assessment practices: a review of international experience. Water Sci Technol: Water Supply 2(2):43–49
Sullivan C (2002) Calculating water poverty index. World Dev 30(7):1195–1211
Tamo Tatiétsé T, Rodriguez M (2001) A method to improve population access to drinking water networks in cities of developing countries. J Water Supply: Res Technol—AQUA 50(1):47–60
UN-HABITAT Urban water and sanitation governance index. http://webworld.unesco.org/water/wwap/wwdr/indicators/pdf/C2_Urban_Water_and_Sanitation_Governance_Index.pdf. Accessed 15 Sept 2019
UNDP, DHA (1994) Disaster Mitigation, 2nd edn.: Disaster management training program. Cambridge Architectural Research Limited, UK
United States Geological Survey (USGS) (2015) National field manual for the collection of water quality data, book 9: handbooks for water resources investigation, US Dept. of the Interior, US Geological Survey, October 2015
United States Environmental Protection Agency (2016) Online source water quality monitoring system for water quality surveillance and response systems. EPA Water Security Division, US
World Health Organization (1976) Surveillance of drinking-water quality. World Health Organization, Geneva
World Health Organization (1993) Guidelines for drinking-water quality: volume 1 recommendations, 2nd edn. World Health Organization, Geneva
World Health Organization (2004) Guidelines for drinking-water quality: volume 1 recommendations, 3rd edn. World Health Organization, Geneva
WHO, UNICEF (2000) Global water supply and sanitation assessment. World Health Organization, Geneva and United Nations Children's Fund, New York
Whittington D, Lauria DT, Mu X (1991) A study of water vending and willingness to pay for water in Onitsha, Nigeria. World Dev 19(2–3):179–198
Woodward A, Hales S, Litidamu N, Phillips D, Martin J (2000) Protecting human health in a changing world: the role of social and economic development. Bull World Health Organ 78:1148–1155

Chapter 9
Managing Rural Drinking Water Supply Across Hydro-climatic Zones of India

9.1 Introduction

There are two factors that threaten the sustainability of rural water supply systems based on groundwater. They are: (1) climate-induced hazards, especially droughts that cause a reduction in water availability in wells; and (2) the competition for water from sectors, especially irrigation. We have discussed how groundwater irrigation impacts the sustainability of rural water supply sources in different situations in Chap. 3. Having said that, the climate itself impacts groundwater demand for irrigation and therefore can impact the sustainability of rural water supply schemes, apart from its direct impact in terms of reduction in water availability in the natural system. For instance, in drought years, irrigation water demand increases owing to reduced moisture availability, and due to reduced supply of water from surface reservoirs (tanks, lakes, and major reservoirs), farmers would increase groundwater pumping if water was available underground. During droughts, since the pumping begins during the monsoon season itself, the water scarcity is felt much before the onset of the monsoon. The opposite can happen in wet years.

Research has shown that the climate-induced water-related hazards that affect the availability of water for drinking water supply in India are highly influenced by hydrological regimes, particularly the average rainfall conditions and the extent to which it varies from year to year (Krishnamurthy and Shukla 2000; Pal and Al-Tabbaa 2011), geohydrological environment (Rawat et al. 2011; Kumar 2014), topography (Kumar et al. 2016; Prokop and Walanus 2017) and climatic conditions (Rajeevan et al. 2008; Guhathakurta et al. 2011). In a typical sub-humid or humid region, the frequency of occurrence of droughts will be less as such regions witness very low variability in annual rainfalls (Kumar et al. 2021). Further, the exposure of the water supply systems to such hazards depends on the characteristics of those systems. In a region underlain by hard rock formations and with steep terrain conditions, the utilizable groundwater recharge and aquifer storage potential will be very low (Jha and Sinha 2009). Hence, any drinking water supply scheme based on wells and bore wells is unlikely to succeed (Kumar et al. 2021).

M. Dinesh Kumar et al., *Drinking Water Security in Rural India*, Water Resources Development and Management, https://doi.org/10.1007/978-981-16-9198-0_9

The climate-induced hazards cannot be mitigated through the system design. The only factor affecting the risk which can be changed through design is the exposure. Designing water supply systems to reduce their exposure to hazards especially droughts and floods call for reducing the negative physical externalities induced by the unique characteristics of the resource, i.e. water—its variability across seasons and years, its stock, its quality and the way quality varies due to various stresses—and the system which provides the service, whether it delivers water inside the dwelling premise or outside; whether it is a closed system (like piped water supply) or an open system (a pond or a well); and the quality of water delivery service.

The characteristics of the resource and the system are governed by several macro-physical and socio-economic variables (UNICEF and IRAP 2017). Some of them are: depth to groundwater table (for WASH); characteristics of the water source (whether a hard rock aquifer with seasonal fluctuations or an alluvial aquifer with minimum fluctuations; whether a perennial river or an ephemeral stream); provision of buffer storage of water in the reservoir in per capita terms; proportion of the households covered by tap water supply; proportion of households having access to modern toilets; and condition of water distribution and delivery system (Kumar et al. 2021).

Reduction of some of these negative physical externalities can either be through changing the source water itself or through changing the condition of the very resource (IRAP, GSDA and UNICEF 2013; Kabir et al. 2016). In a semi-arid or arid plain with hard rock aquifers, the option will be to either go for a regional water supply scheme based on a distant reservoir or make sure that the aquifers get continuous replenishment through recharge from irrigation water, especially during the summer months. If the water supply source is highly dependable, yet people depend on stand posts, then improving the access to the provision of household tap connections can reduce the problems associated with water contamination during collection and transport (Kumar et al. 2021).

This chapter analyses the macro-level conditions that exist in India vis-à-vis the key attributes that have bearing on the type of water supply schemes that would be sustainable in a rural area. The attributes considered for analysis are: surface hydrology; topography; geology and geohydrology; groundwater chemistry; and degree of over-exploitation of groundwater. The regions in India where groundwater-based schemes are unlikely to be sustainable are then identified and the reasons for the same are explained. Following this, strategies for sustainable rural water supply in those regions from surface water resources are evolved, based on consideration of surface hydrology and topography. The chapter also discusses the key institutional and policy reforms related to water that are required to manage water resources and drinking water supplies on a sustainable basis under climate extremes.

9.2 India's Surface Hydrology

It is an established fact that more than two-thirds of India's surface water resources are in the Ganga–Brahmaputra–Meghna River basins (GOI 1999). A quick analysis of the estimates of dependable yield of 20 river basins (CWC 2017) shows that the runoff rates (expressed in MCM per km^2 of the drainage area) show sharp differences in runoff rates between river basins. The lowest is in the Luni river basin of western Rajasthan. The highest is in Barak and other river basins of the north east (CWC 2017).

Rainfall and climate are the two most important factors causing variation in water availability across India's river basins. Though there are other factors such as soil type and land cover which influence surface runoff generation, their effects on runoff generation are not discussed here as the spatial variation in such features observed between river basins is not as distinct as the variation in rainfall and climate in India (source: based on CWC 2017 and Pisharoty 1990), when it comes to impact on runoff. In India, the regions which experience very low to low rainfalls witness very arid climates and vice versa. The north-eastern region which receives more than 3,000 mm of annual precipitation has an annual potential evaporation of around 1,500 mm, whereas western Rajasthan which receives around 200 mm of mean annual precipitation has an annual potential evaporation of more than 3,500 mm. Such unique patterns in the spatial variation of rainfall and climate result in drastic variation in surface hydrology (Kumar et al. 2021).

The excessively high rainfall region of north-east has a cold and humid climate. Due to this reason, the soil moisture depletion is very low. It is evident from the fact that the average reference evapotranspiration for the region (estimated for Umiam in Meghalaya using the Penman method) is only 2.85 mm/day. The ET_0 for the southwest monsoon season is 2.64 mm (Rao et al. 2012). Hence, a major share of the rainfall from each rainfall event gets converted into surface runoff, with very little initial absorption of precipitation into the soil profile. The hard terrain with very thin soil cover also hinders infiltration of the incident rains. Conversely, in the hyper arid region of western Rajasthan, a very large proportion of the annual precipitation infiltrates into the ground. The potential evaporation for Jaisalmer is 2,590 mm, whereas that for Pali is 1990 mm. The average reference evapotranspiration for Jodhpur (falling in Luni basin) is 8.15 mm per day, estimated as per the Penman method. Even during the southwest monsoon season, the average daily value is as high as 8.78 mm (Rao et al. 2012). Due to the high aridity, the process of depletion of soil moisture in this region is very rapid in the region.

As a result, Barak River has a dependable yield of 1.55 MCM per sq. km, while the Luni river basin has a dependable yield of a mere 0.00687 MCM per km^2. The differences are vast. The total yield of the Barak basin, with a catchment area of 47,440 km^2 is 68,580 MCM, whereas that of the Luni basin with a drainage area of 69,000 km^2 has a dependable yield of a mere 474 MCM.

Figure 9.1 shows the rainfall and runoff rates for 20 major river systems of India. Significant differences in the runoff–rainfall ratio can be seen in river basins in the country. In the case of the Brahmaputra basin, 68% of the precipitation is available as surface runoff. In the case of west-flowing rivers (south of Tapi), nearly 66% of the rainfall is converted into runoff. But in the case of Pennar, only 13% of the rainfall is

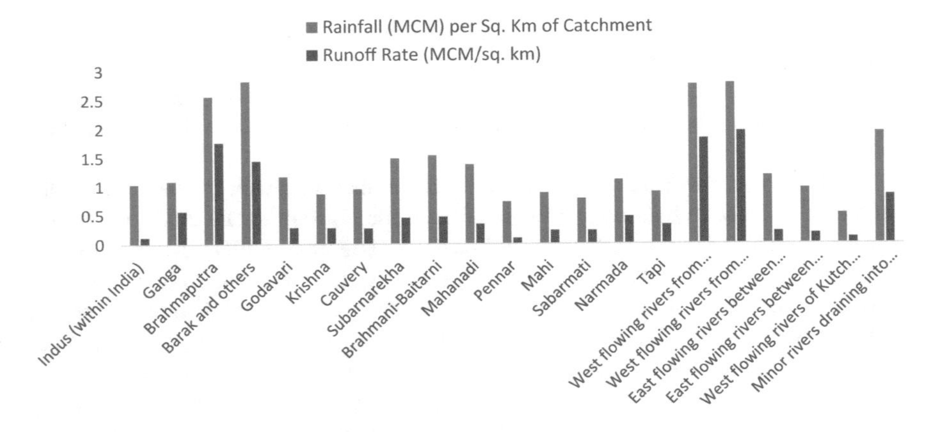

Fig. 9.1 Average rainfall and runoff rates of major river basins

available as runoff.[1] As a study of river catchments in the Western Ghats has shown (NIH 1999), the baseflow contribution to total runoff is significant in basins with steep catchment slopes for the low incidence of rainfall but decreases drastically as rainfall increases. It is also understood that when the groundwater depletes with a steep fall in water table, the interaction between surface water and groundwater can reverse (Winter et al. 1998).

One important concern with surface water availability in India is the high inter-annual variability in the runoff in regions having low mean annual rainfall and high aridity. In such regions, the runoff generation in low rainfall (or dry) years is disproportionately lower than the extent of decline in rainfall from the normal value, and runoff generation in high rainfall (or excessively wet) years is disproportionately higher than the extent of increase in rainfall. Whereas in the humid and sub-humid regions with high mean annual rainfall, the inter-annual variability in surface runoff is very low.

9.3 India's Groundwater Situation

India has heterogeneous geohydrological conditions. The country is underlain by hard rock formations for two-thirds of its geographical area, with basalt, crystalline formations, and limestones. The entire Deccan plateau is underlain by basaltic and crystalline rock formations. The Deccan trap basalt has weathered rocks, underlain by fractured rocks and bed rocks. There are also semi-consolidated formations of sedimentary origin (mostly sandstones) mostly found in western Rajasthan and

[1] The catchment yield considered here also includes the baseflow (groundwater outflow into streams), which is influenced heavily by the formation characteristics and terrain conditions that govern the groundwater flow gradient.

Kachchh. Unconsolidated and deep alluvial deposits are found in the Indo-Gangetic plains—extending from Punjab in the north-west to West Bengal in the east—and in the Cambay basin in Gujarat. Figure 9.2 shows the geohydrological map of India.

The formation characteristics and rainfall together determine groundwater availability in different aquifers of India. Given the wide spatial variation in rainfall and its pattern of occurrence, recharge rates also vary even within the same geological

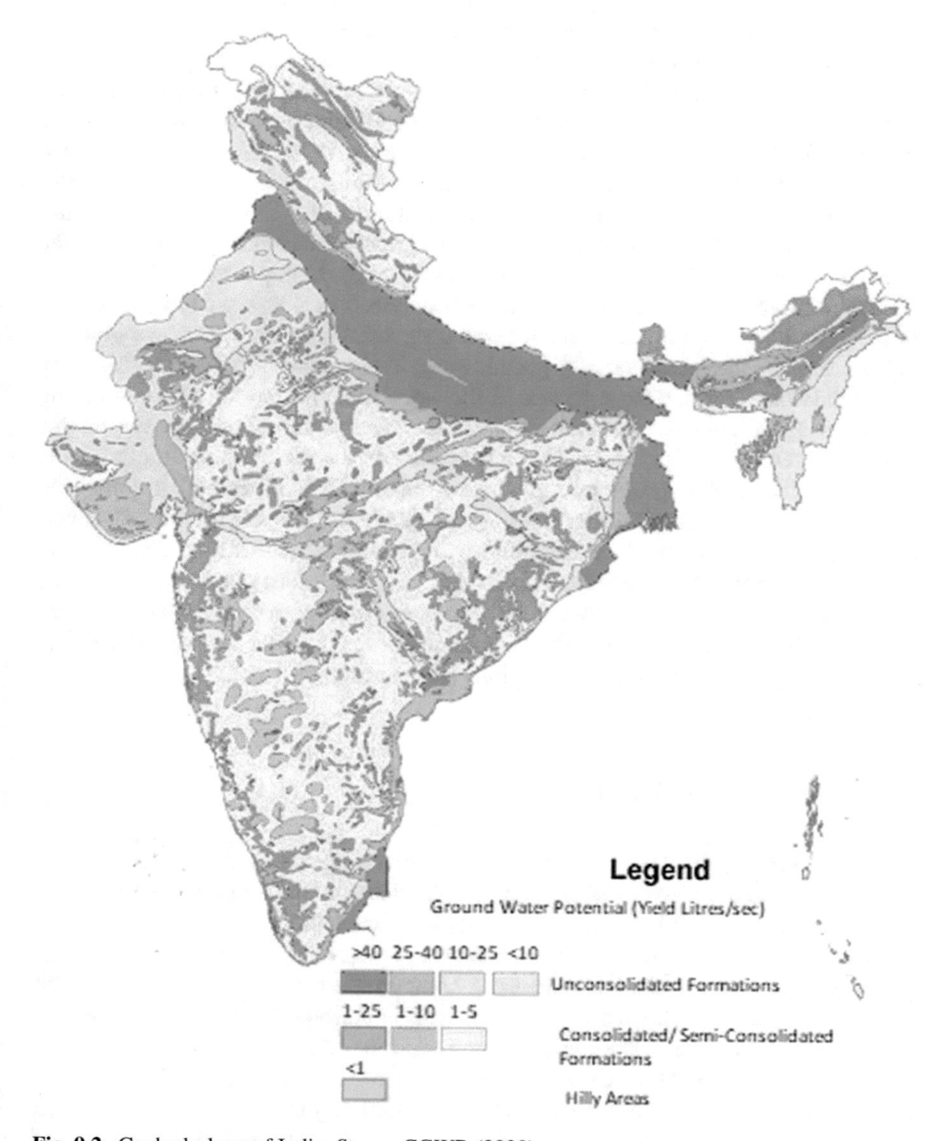

Fig. 9.2 Geohydrology of India. *Source* CGWB (2020)

setting. The deep alluvial deposits are very good aquifers having very high porosity. They have static groundwater, in addition to the dynamic resource. Sedimentary formations are the second best in terms of yield, followed by weathered basalt. The crystalline rock formations, having no primary porosity and with water occurring only in fissures and cracks, have low yields.

Depth to groundwater varies from region to region depending on the rainfall, climatic conditions and geological environment. The depth to groundwater in the homogenous alluvial aquifers varies depending on the rainfall and climate, with depth to water level increasing with reducing rainfall and increasing aridity. In the hard rock aquifers, though depth to water levels is generally not very large, sharp seasonal variation in water level is encountered with increasing draft after the monsoon season. The inter-seasonal fluctuation in groundwater is very large in the hard rock formations. During the monsoon, the water level rises as precipitation increases, and aquifer saturation is common in large parts of peninsular, central India, and east-central India with hard rock formations. The water-level fluctuations during monsoon in the alluvial formations are generally very low, owing to the high porosity.

Groundwater occurrence is also heavily influenced by geomorphological conditions. In regions with mountainous or undulating terrain, groundwater flow gradient is high. Hence, the lateral flow of groundwater is significant. In areas with sloppy and undulating terrain, a large portion of the water that infiltrates the ground during monsoon is available as 'outflows' and joins the streams during the lean season due to the difference in water level between the river and the surrounding aquifer.

Due to the unique geohydrological environment and the rainfall regime and climatic conditions, groundwater mining and steady decline in groundwater levels have been occurring in the alluvial aquifers with a semi-arid and arid climate, covering north and central Gujarat, central Punjab, and western Rajasthan. In the hard rock areas, the problem of groundwater over-draft is confined to very small pockets. However, the problem of well failures, seasonal drops in water levels, and summer water scarcity are rampant in these regions (Kumar et al. 2012, 2021).

9.4 India's Topography

India's topography displays vast heterogeneity from long coastal plains to vast plains of the Indo-Gangetic River system to plateau to desert regions to undulating terrain to mountainous areas (Fig. 9.3). The coastal plains are in the states of Gujarat, Maharashtra, West Bengal, Odisha, Andhra Pradesh, Tamil Nadu, Kerala, and Karnataka. The Gangetic alluvial plans extend from Uttar Pradesh to Bihar to West Bengal (Kumar et al. 2021). This region is dotted with numerous wetlands in the form of floodplains of the rivers, lakes, and ponds. Assam has a large area under the floodplains of the Brahmaputra River. The Deccan plateau extends from Saurashtra in Gujarat to large areas of Maharashtra (Marathwada), South eastern part of Madhya Pradesh, parts of Andhra Pradesh, Telangana and Karnataka, and some parts of Tamil Nadu. The other smaller plateaus are Marwar-Mewar plateau, Chhotanagpur

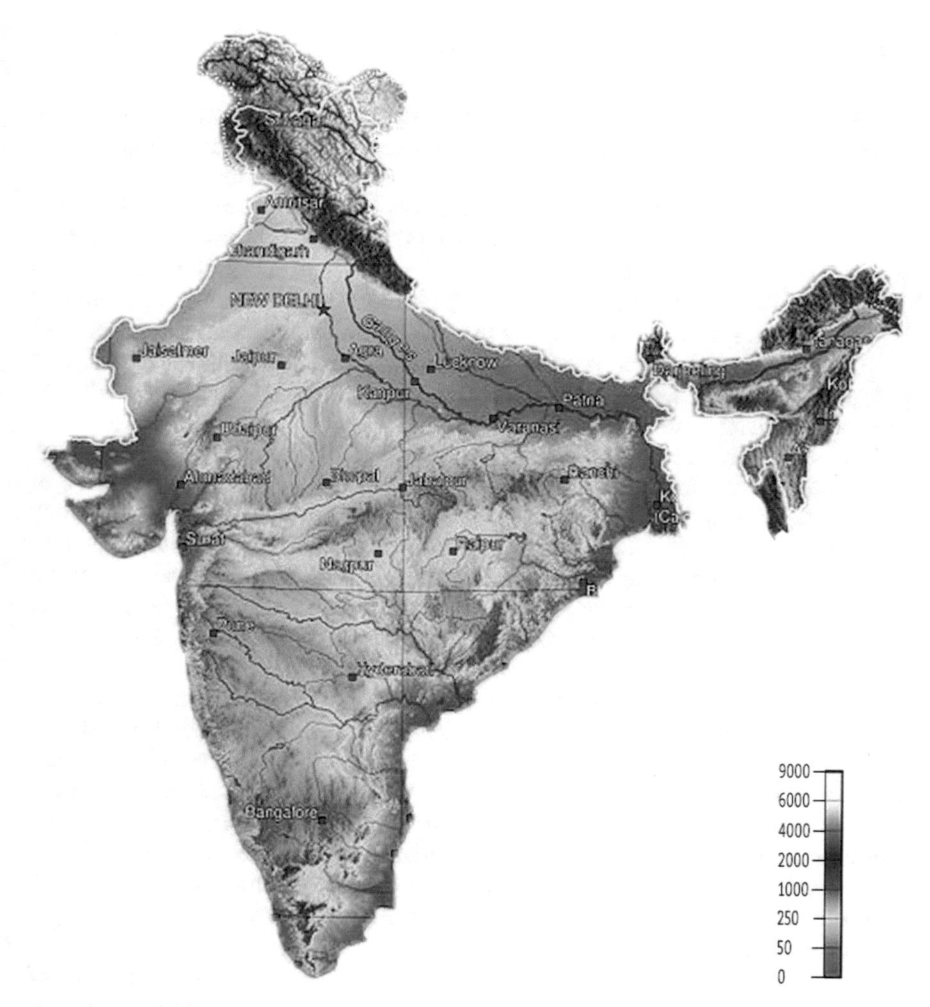

Fig. 9.3 Topography of India

plateau, and Bundelkhand plateau. The Western Ghats run through the southern part of Gujarat through western Maharashtra to Goa, and western Karnataka to Kerala. The Eastern Ghats run through Tamil Nadu, Andhra, and Odisha (Kumar et al. 2021). Both the Western Ghats and Eastern Ghats are well drained by several small rivers with steep channel slopes. Parts of Punjab and the whole of Haryana are part of alluvial plains. The Aravalli mountain ranges run in the south-east to north-west direction in Rajasthan and extends up to Haryana. The Vindhyan and Satpura ranges run almost parallelly through Madhya Pradesh in the east–west direction. The sub-Himalayan ranges extend over Jammu & Kashmir, Himachal Pradesh, and many states of the north-east.

9.5 Strategies for Sustainable Rural Water Supply

Research carried out in Rajasthan on climate-induced risk in WASH has also shown that the places that secure highly dependable sources of surface water through imports (e.g. Jaisalmer district of western Rajasthan) experience very low exposure of the water supply systems to droughts, in spite of being very arid and naturally water-scarce. The basic reason is that the water supplied through the piped water supply schemes there is drawn from the IGNP canal which gets its supplies from a very dependable source, i.e. the Sutlej River which is fed on the glaciers of Himachal Pradesh. The research has also shown that the households which have access to a dependable source of treated water in their dwelling premise through tap connection pose very low public health risks associated with poor water supply and sanitation services. (Hemani et al. 2021).

That said, only 18% of the rural households in India have access to *tap water*, i.e. treated water supplied through pipes to the dwelling. A large percentage of the population receives piped water supply, but they do not have household tap connections. Also, the water is untreated in most cases. Three factors hinder people's access to '*tap water*'. They are: economic conditions of the HHs; quality of water supplied; and lack of assurance about the ability of the scheme to supply water throughout the year. The second and third factors are linked to the characteristics of the resource.

Nearly 80–85% of the rural water supply in India is dependent on groundwater. The low gestation period and the low maintenance requirements had attracted both the water supply agencies and the rural communities towards groundwater-based schemes. Small cities and towns began building water supply schemes with water from local rivers, lakes and tanks, then from local groundwater, and then finally moved to import water from distant reservoirs and rivers as local sources became highly inadequate. With the over-exploitation of groundwater for irrigation, and with increasing chemical contamination, schemes based on groundwater are increasingly becoming unsustainable in rural areas.

The problem is more acute in the hard rock regions where the dependability of groundwater is quite poor and where individual village-based drinking water schemes based on open wells and bore wells are very common. In Maharashtra, for instance, the extent of dependence on groundwater sources for rural water supply ranges from a lowest of 25% to a highest of 97%. While these aquifers mainly obtain their annual replenishment from monsoon rainfall, the poor dependability of groundwater in these regions is because of the quantum of rainfall and pre-monsoon depth to water table, the two factors influencing the water table fluctuation and recharge during monsoon. When the rainfall is low, the recharge is less. When the pre-monsoon depth to water table is low, then again recharge suffers. The key step towards providing tap water to every HH in such situations is to change the source to the more dependable surface water (in areas where SW availability is good) (Kumar et al. 2021).

Figure 9.8 shows the geographical extent of the problem of groundwater depletion in India, expressed in terms of depth to water levels (pre monsoon) in 2019 (CGWB

2020). We have seen in Chap. 7 that large regions in India face groundwater quality problems. The problem of over-exploitation is mostly confined to the alluvial areas of north Gujarat, western Rajasthan, Punjab and very small pockets in peninsular India (in Tamil Nadu, Karnataka, Andhra Pradesh, and Telangana). The problem of season depletion of groundwater, which results in acute summer water scarcity, is prevalent in the entire hard rock region of India—Telangana, Chhattisgarh, Karnataka, and large parts of Andhra Pradesh, Tamil Nadu, Madhya Pradesh, Maharashtra, and Odisha.

The resource problem in these hard rock regions that experience seasonal depletion is due to its excessive withdrawal for irrigation (with low to medium rainfall, high aridity and a large amount of arable land that needs irrigation) during the winter, which leads to drying up of the aquifer by summer. In drought years, the pumping of groundwater for irrigation can start during the monsoon season itself, resulting in faster depletion.

The shift to surface sources for drinking water supply is needed in regions where groundwater is either over-exploited or its supply is unreliable or is chemically contaminated. There are two advantages of shifting to surface reservoirs for drinking water supply from the point of view of climate risk aversion. First: hydrological planning of the scheme is possible, thereby the size of the reservoir and catchment can be decided on the basis of the estimated water demand in the service area. Conversely, based on the amount of water in the reservoir, it is possible to arrive at the size of the service area and population that can be served. Second: physical allocation of water from the stock of water in the reservoir is possible and the problem of competition for water between sectors, which is confronted with in the case of water from underground sources, does not exist.

Therefore, the need for tapping surface water for domestic supplies exists in the following regions: (i) Saurashtra, Kachchh, and eastern parts of Gujarat (hard rock) and north Gujarat (TDS and fluoride); (ii) Rajasthan—western Rajasthan, due to high salinity and fluorides, and the rest due to hard rock formations; (iii) the arsenic affected areas of West Bengal and Bihar; (iv) the entire state of Karnataka, except coastal regions; the entire state of Telangana; AP, barring the costal deltaic region; the entire state of Maharashtra, excluding the Kongan region; (v) the entire Madhya Pradesh state, except the alluvial belt of central Narmada valley; and the entire state of Chhattisgarh; (vi) the entire state of Odisha except the coastal alluvial belt (due to hard rock formations); (vii) the high ranges and midlands of Kerala that have hard strata and poor groundwater potential; (viii) the alluvial areas of Punjab and Haryana that do not receive canal irrigation; and, (ix) the entire hilly region of north eastern India, and J&K region.

The question comes as to where we can find surplus water for augmenting the existing or developing new water supply schemes. The basins such as Brahmaputra, Ganges, Mahanadi, Mahi, Tapi, Godavari, Brahmani-Baitarani, WFR south of Tapi and EFR between Mahanadi and Pennar are actually '*water-surplus*' as the demand for water in agriculture is low against high flows; the Narmada basin is neither surplus nor stressed; Sabarmati, Pennar, the west flowing rivers of Kachchh and Saurashtra, Krishna, Cauvery and east flowing rivers south of Pennar are either water scarce or absolutely water scarce, as agricultural water demand is very high in these basins, against low flows. The question at the next level is how much of these surface water resources are stored, against what we require to meet the domestic water needs. Further, we need to know how much additional storage can be created in the future. Let us examine these questions.

We need 39,370 MCM (39.37 BCM) of water annually for domestic water supply in rural areas if we consider an average per capita daily water demand of 150 L, and for 80% of the total country's rural population (projected to be around 900 million by 2021). However, from this, the population of UP, Bihar, West Bengal and Assam can be excluded. The reason is that there is plenty of groundwater in the Gangetic alluvial plains covering UP, Bihar, West Bengal, and Assam, which accounted for 34% of the country's population in 2011. The water supply schemes in this region do not face any threat from agriculture as groundwater draft is unlikely to increase there. Schemes based on tube wells tapping water from shallow aquifers/deep aquifers will be sustainable in this region, except for a few pockets that experience the problem of arsenic. In the arsenic-affected areas of WB and Bihar, schemes based on large ponds and lakes shall be planned, with rapid sand filters (Kumar et al. 2021). Hence, the total amount of surface water required will be only 66% of the estimate we have provided earlier, i.e. 25.98 BCM.

The live storage capacity of reservoirs in India stands at around 230 BCM. Figure 9.4 shows the live storage capacity of large reservoirs in major Indian states. The effective water utilization capacity of these reservoirs could be at least 20–25% more as many of them divert water during the season in which the inflow is received. Hence, the available reservoir storage capacity is at least 5–6 times higher than the total volume of water required. The entire surface runoff in Krishna, Cauvery, Sabarmati, Pennar, Luni, the west-flowing rivers of Kachchh and Saurashtra, and east-flowing rivers south of Pennar is tapped through M&M reservoirs for irrigation and domestic use. Further augmentation or reallocation is not possible due to conflicts between sectors and states. But there is still scope for storage enhancement/diversion

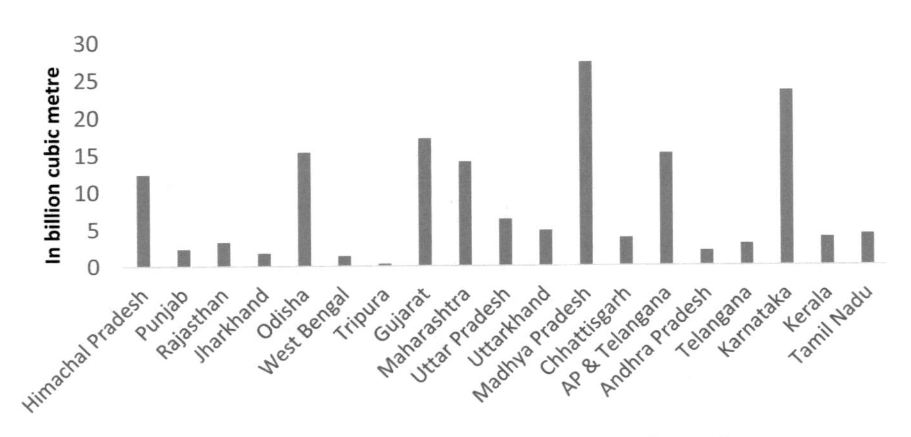

Fig. 9.4 Live storage of large dams in major Indian states (billion cubic metre)

of water in many basins—Mahanadi, Brahmani-Baitarani, Narmada, Godavari, Tapi, WFR South of Tapi and EFR between Mahanadi and Pennar.

Scope exists for the transfer of surplus water from Sutlej River during wet years through the Indira Gandhi Nehar Project (IGNP). Though the Brahmaputra basin, and to some extent the Ganges, have a large quantum of water, there is no feasibility due to topography, surface morphology or heavy sediment transport. Water transfer possibilities exist between Godavari and Pennar and then further down south to Chennai. Possibilities also exist for the transfer of water from South Gujarat (Karjan and Tapi to Narmada and further north) to north Gujarat. Kerala river basins produce an annual runoff of 70.30 BCM, with high dependability. The gross reservoir storage created is only 6.90 BCM. Possibilities do exist for the transfer of water from Kerala to Tamil Nadu.

Having got some insights about the storage of surface water available annually and the scope for augmenting the storage in the future, let us look at the distribution of the existing storage facilities. India has around 5,101 large dams (height > 15 m or storage > 1.0 MCM). Around 36% of them are in Maharashtra (around 1,821). There are around 61 very large reservoirs with a capacity of more than 1,000 MCM or height > 100 m. The large dams in Madhya Pradesh, Gujarat, and Karnataka have the highest storage.

The question now is 'what experience does exist in India for bulk transfer of water from large reservoirs for meeting domestic water demands that are highly distributed?'. Water allocation for Gujarat from the Narmada basin done by the NWDT (1979) considered the drinking water requirement of the state. The idea of bulk water transfer to Saurashtra and Kachchh emerged sometime in 1999, soon after a devastating drought, when the water harvesting initiatives failed to provide any relief to these regions. The key to the success of this initiative lies in the sophisticated water infrastructure consisting of large diameter pipes running several thousands of kilometres, with distribution lines taking water to several thousands of villages in Saurashtra, Kachchh, and eastern and northern parts of Gujarat.

That said, we will now examine the specific interventions possible in different regions for improving the water supply. As far as possible, we will do this river basin-wise.

9.5.1 Water Supply from Mahanadi River Basin

Nearly 60% of the geographical area of Chhattisgarh state, which falls in the Mahanadi River Basin, can be covered by water from new reservoirs/weirs and reallocating water from the existing ones in that basin. The rest of the area of the state can be covered by water from the Godavari. Large areas in western Odisha can be covered by regional water supply schemes built around reservoirs in the Brahmani-Baitarani basin, and Hirakud reservoir in Mahanadi, which is on the border of Chhattisgarh and Odisha.

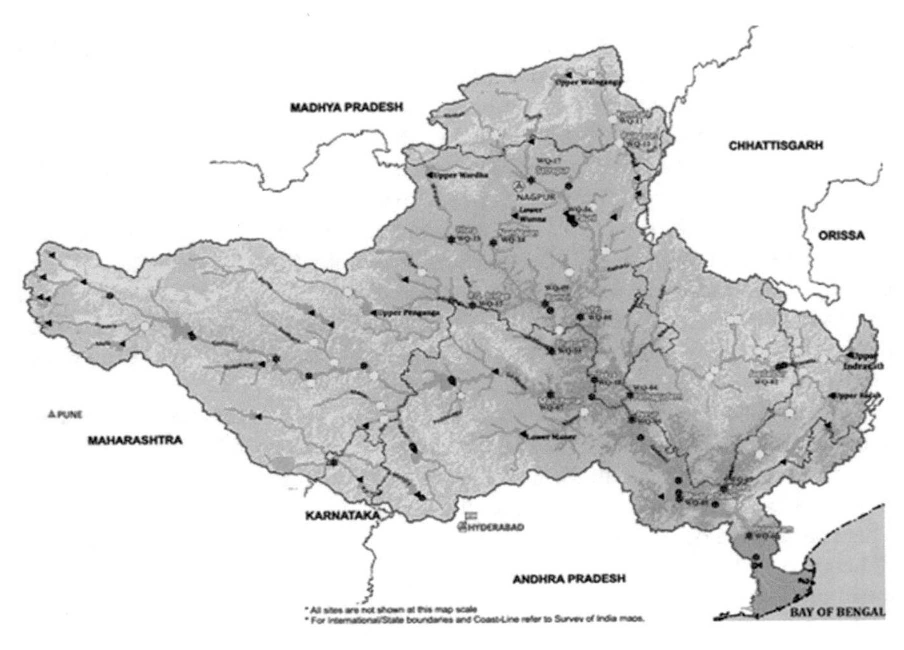

Fig. 9.5 Map of Godavari river basin

9.5.2 Water Supply from Godavari River Basin

Godavari river basin is presented in Fig. 9.5. Water supply for the entire Rayalaseema region, which is one of the most water-scarce regions of south India, can be provided through the import of water from the Godavari to the Krishna basin and then to the reservoirs of Pennar basin through water transfer links. Water lifted from Godavari will be able to provide drinking water supply to the villages, towns and cities of Telangana, as infrastructure is already established for lifting large volumes of water (around 138 TMC or 5,040 MCM per annum over a period of 90 days) from the river under the Kalpeshwar project. The tanks in the region can be used as intermediate storage reservoirs for storing the diverted 'monsoon flood water'. Water from the upper catchments of the Godavari River Basin in Maharashtra can be harnessed through medium reservoirs for distribution through pipelines in the Vidarbha region.

9.5.3 Water from Sutlej/Beas Rivers

Surplus water from Beas/Sutlej rivers can be diverted to the Indira Gandhi canal traversing through the deserts of western Rajasthan for providing piped water supply to new districts in western Rajasthan (other than Jaisalmer, Barmer, Ganganagar and Churu). Intermediate storage systems would be required for storing the water from the canal system as flow is not perennial. While Punjab has one of the most well-developed irrigation systems with canal networks and tube wells that irrigate around 95% of the cultivated area of the state, drinking water supply systems are still not capable of securing freshwater in many areas. The state has nearly 1.9 million rural households with water connections, but the water supply is dependent on groundwater. Around 1,100 villages in Punjab experience problems of high fluoride and arsenic in groundwater. Groundwater in many villages faces contamination from Uranium too. The state has a very good network of canals supplying surface water for irrigation from the Bhakra-Nangal project, on the Sutlej River. Drinking water supply schemes can be built around this water—it is free from chemical contaminants. Excess flows from Sutlej can be diverted and stored in intermediate storage tanks.

9.5.4 Water from Sardar Sarovar Reservoir

The Sardar Sarovar reservoir located in Gujarat in the Narmada River Basin has a gross storage capacity of 10,800 MCM. Water from the Sardar Sarovar Project is serving 10,500 villages in Gujarat (covering Saurashtra, Kachchh and eastern Gujarat) and villages of Jhalore in Rajasthan. The main canal of Sardar Sarovar Project (SSP), having a length of 458 km, goes up to south western Rajasthan.

In addition to the SSP, twenty-eight large and 125 medium reservoirs are proposed under the Master Plan for Narmada Valley Development in Madhya Pradesh. Regional water supply schemes can be planned around these reservoirs to cater to areas inside the basin and neighbouring areas.

9.5.5 Water Supply from the Tributaries of Ganges

A small portion of the Chambal River basin is in Rajasthan and a large portion is in Madhya Pradesh. The northern parts of Madhya Pradesh can be served by reservoirs that exist in the tributaries of Ganges River and Yamuna River (Chambal, Ken, Betwa, Sone, Tons, Paisuni and Baidhan, and Jamni) passing through the state. The villages/towns of southern and south-eastern Rajasthan can be served by a few reservoirs in the Chambal basin and small reservoirs/lakes in the local catchments (in Udaipur, Dungarpur, Pratapgarh, etc.).

9.5.6 Water Supply for River Basins in Karnataka and Tamil Nadu

Figure 9.6 shows the map of Karnataka with the major river basins. A small proportion of the water from the reservoirs in the Bhima, Tungabhadra, and Krishna rivers can be allocated for water supply in the northern, central, and north eastern parts of the state. Water from reservoirs in the Cauvery basin is to be diverted for the southern and south-eastern parts. Water from reservoirs in the Western Ghats in Karnataka can be taken to south central and western parts of the state. Taking water from the reservoirs in the Western Ghats and diverting to south-central and western parts would require lift. The coastal areas do not require surface water.

Figure 9.7 shows the map of Tamil Nadu with the major river basins. Majority of the geographical area of Tamil Nadu is underlain by crystalline rocks. Water shortage is severe in the Madurai region and Chennai. Some water from reservoirs of the basins in the region is to be reallocated for domestic use in towns and villages of Madurai. Chennai can get water from the Telugu Ganga project from Godavari, which gets transferred to Krishna and then Pennar basins before moving further down to the south. Water transfer possibilities from Kerala is to be explored for inter-state and other basins, for serving areas within the Cauvery basin (Fig. 9.8).

9.5.7 Water Supply for Kerala and the North Eastern Hill States

Kerala on the southern tip of India is unique in many ways. The state is first from the top in terms of progress in human development. The state's human development index is 0.795. Nearly 3.5–4.0 million households in Kerala have access to mini water supply schemes with household tap connections. Open wells in the homestead are the sources. Water scarcity hits the rural and urban areas during summer (April–May) with no rains and excessive pumping of well water for irrigating plantation crops such as coconut, araca nut, and banana. Bacteriological contamination from septic tanks is emerging as a major issue in the state. Water from existing reservoirs and a few new reservoirs in the WFRs of Western Ghat region (in Kerala) is to be supplied to villages and towns in the high ranges and midlands of Kerala.

The hills of the northeast receive rainfall above 3,000 mm, yet face acute drinking water shortages during summer months as wells and springs dry up; people there depend on local streams for domestic water supply. Small dams can be built to store runoff from the local catchments of size 10–15 km^2. The runoff (20–30 MCM) will be sufficient for 2–3 lac people. A geotechnical investigation will be required to ensure slope stability and safety of the dam. Direct lifting of water using pumps is required to serve the habitations upstream of the reservoirs. Spring water can be diverted by gravity and stored in underground reservoirs at the hamlet level.

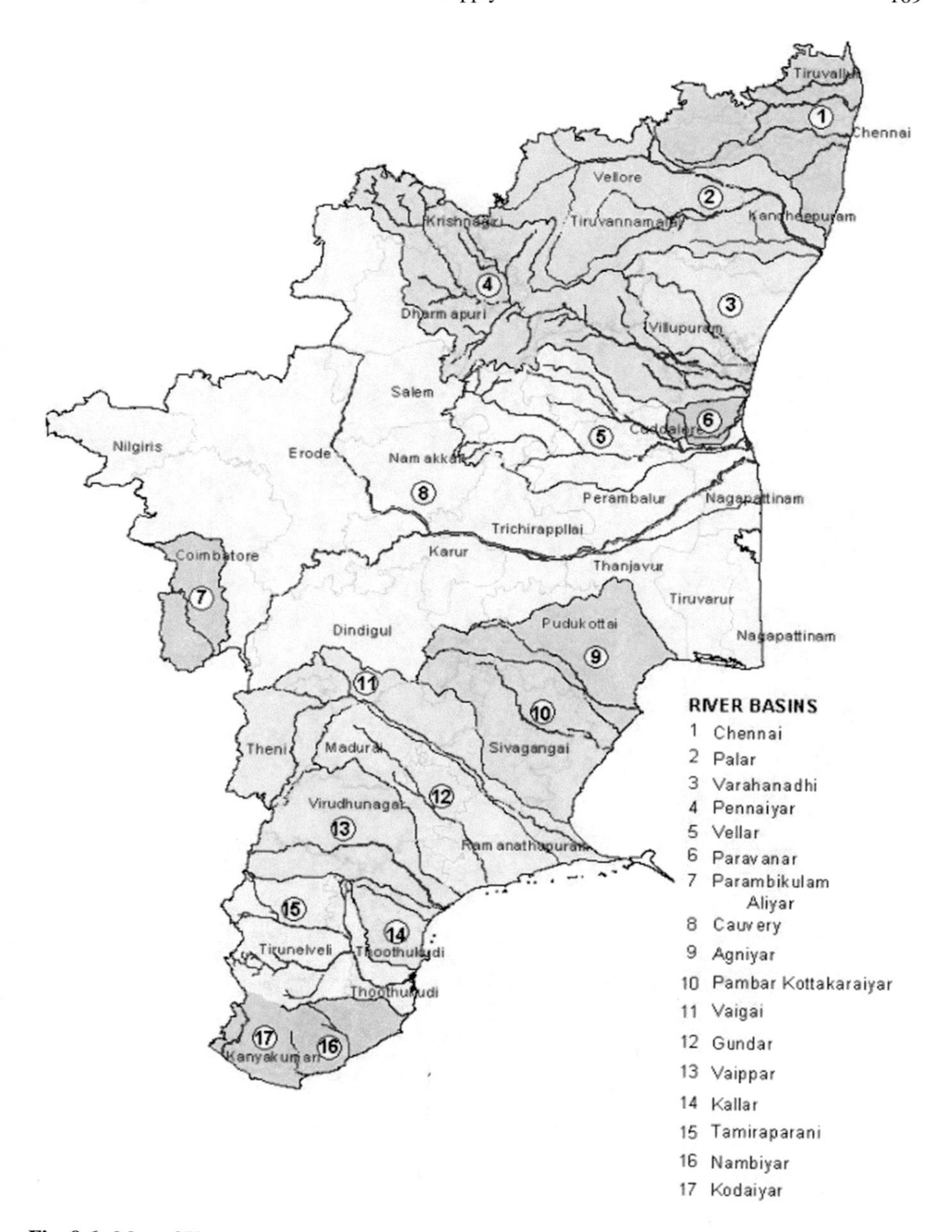

Fig. 9.6 Map of Karnataka showing the major river basins in the state

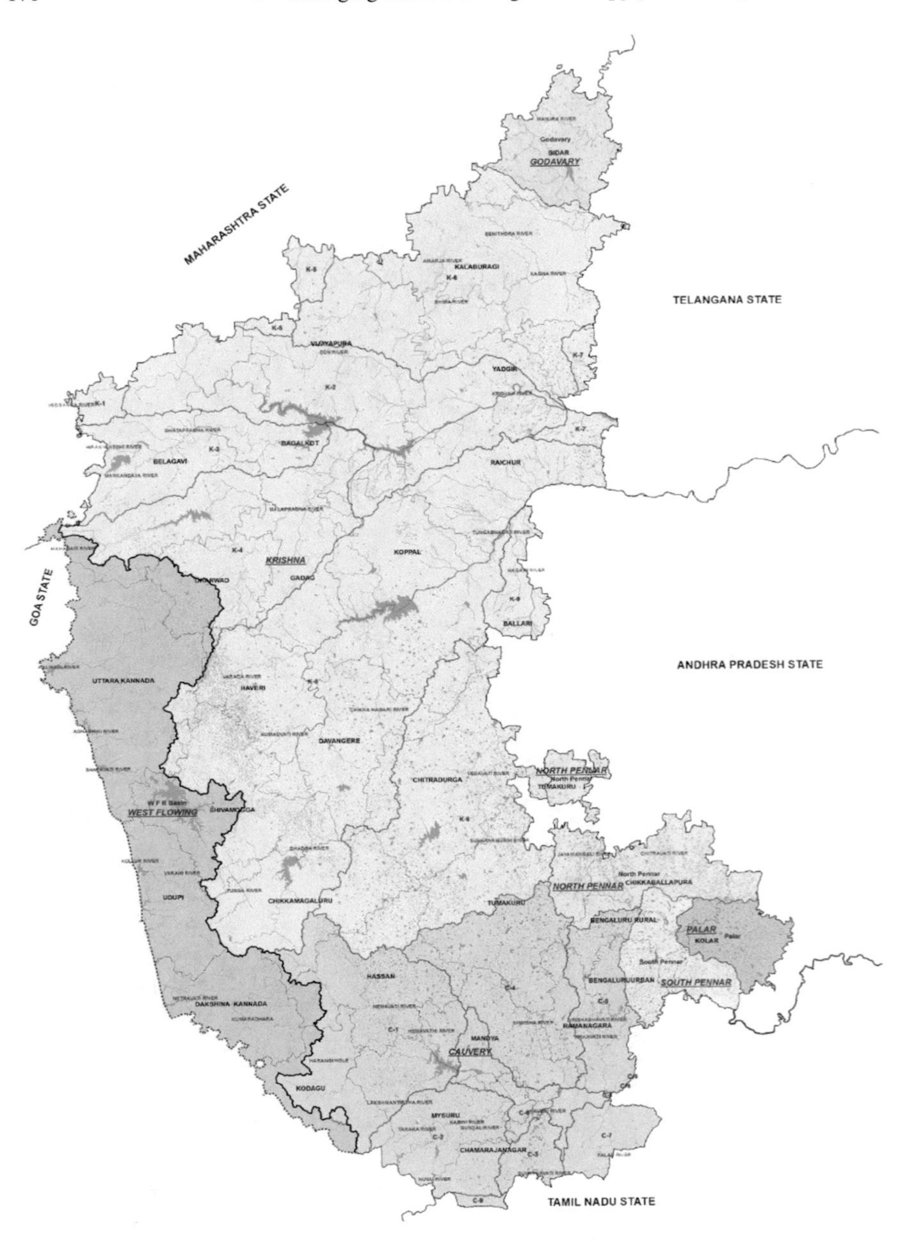

Fig. 9.7 Map of Tamil Nadu showing the major river basins in the state

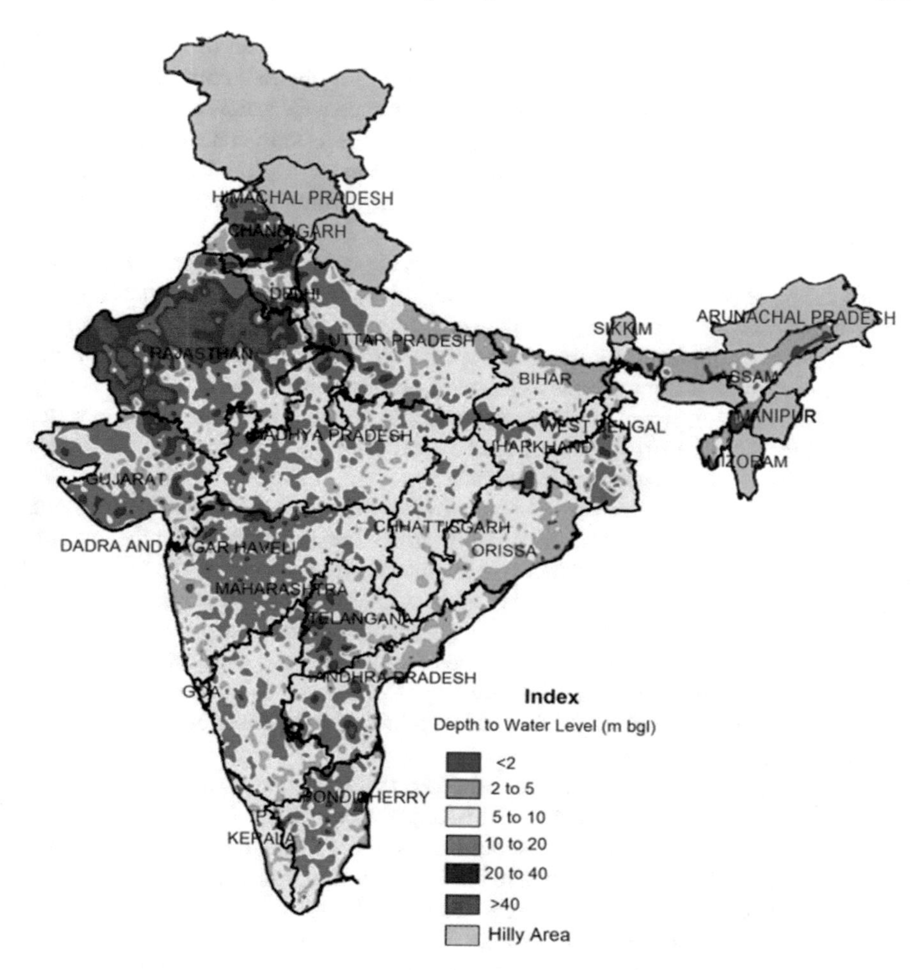

Fig. 9.8 Map showing depth to groundwater levels in India (pre monsoon 2019)

9.6 Institutional and Policy Choices for Improving Drinking Water Security

9.6.1 Institutional Strengthening

One major strategy to improve drinking water security in large parts of India, especially the naturally water-scarce regions, is to implement inter-basin water transfer projects (also see Kumar 2018; Kumar et al. 2021). This would require consensus between the donor states and the states that receive the water through such transfers. Under the current law, water is a state subject and the states can exercise full control over the utilization of their water resources, except in the case of inter-state river basins where the Union government has a say in the matters concerning sharing of water and also benefit-sharing. The past experience with water sharing with such water projects involving inter-basin transfers is that the 'donor' state fiercely opposes such ideas based on the fear of millions of people in rural areas losing their livelihoods based on water (due to adverse effects on agriculture, fisheries, drinking water supplies).

For inter-basin water transfers to be a reality, the current legal regime with regard to utilization of water resources within the administrative jurisdiction of states, which have abundant water resources, will have to change for them to be under the purview of national laws. This can only enable speedy decisions for the development and utilization of these water resources. However, political consensus is likely to emerge in the future between potential 'donor states' and receiving' states on sharing of water. As noted by Kumar (2018), we can also expect a greater application of economic principles in water management in the future.

This central law can provide for evolving mechanisms for sharing of benefits from such inter-basin water transfer projects, in order to compensate for the losses to the donor states. The estimated increase in agricultural production due to increased utilization of water in the water-scarce region or the estimated power generation due to the building of a hydropower dam or the estimated increase in drinking water supply benefits or all of them (in the case of multi-purpose projects) can be used as the basis for benefit sharing. The new law can have provisions to set up judicial panels to adjudicate on water sharing and payment of compensation, with technical expert committees to evaluate benefits.

As regards the quality of water for drinking, it is important that the water supply agencies improve the surveillance of drinking water sources for preventing water contamination. Such surveillance should be based on risk assessment. The task of assessing the health risk associated with water contamination is complex for groundwater-based sources. In Chap. 8, we have discussed how the drinking water quality surveillance index (DWQSI) can be used to assess the public health risk associated with water contamination as a tool to decide on the surveillance strategy for different regions. Though the computation of this index would require a lot of data on the physical environment (hydrology, geohydrology and climate), socioeconomic conditions and institutional setup, a rapid assessment of the macro-level situation suggests that such risks through water contamination would be high in the very high and excessively rainfall areas of eastern Gangetic plains and Brahmaputra plains, and coastal plains of Kerala, Odisha, Karnataka, Andhra Pradesh, and Maharashtra, having humid to sub-humid climate and shallow groundwater, and can be due to nitrate contamination (Kraft and Stites 2003) and microbial contamination of groundwater (source: based on Chap. 7, this book).

In these regions, the state water supply agencies should improve their capabilities to protect the groundwater-based drinking water sources through a well-designed surveillance programme that involves frequent monitoring of the quality of source water, and quick dissemination of the results to the local institutions that manage the drinking water sources at the village level.

9.6.2 Changes in Policy Framework

9.6.2.1 Inter-sectoral Water Allocation Policies

The National Water Policy (2012) envisions that domestic water supply gets priority over other sectors of water use in water allocation decisions. The policy also emphasizes using economic principles for the pricing of water. But, the use of the economic principles for the pricing of water would be inappropriate without prior allocation of water for various sectors of water use. The reason is that the economic principles suggest that marginal returns from the use of water should form the basis for fixing water prices for ensuring affordability. Since the marginal returns are very high for manufacturing, the affordability in this sector would be generally high. Under such circumstances, the blanket use of this policy can lead to over-allocation of water from river basins to the manufacturing sector and other sectors where water is used for economic activities, depriving the priority use sectors (Kumar et al. 2021).

Hence, it is important that from the total renewable surface water in the basin, which includes what is already appropriated through reservoirs and diversion systems and the "un-committed flows", a portion is first allocated to the high priority sector of domestic water use in rural and urban areas, based on the actual requirements of the respective sectors. Further allocation from the balance can be made to other competing use sectors such as irrigation and industry. The demand for water for drinking and domestic uses can be estimated on the basis of the population size and the realistic norms of per capita water supplies. The practice followed at present is that when there is a drought, the Collector of the concerned district orders freezing of the public reservoirs in the district, and with such an order the water resources department is prevented from releasing water for irrigation. Such an approach is ad hoc and can induce unnecessary stress on the irrigation agency and the farmers as the water required for domestic water supplies could be quite a small proportion of the total water available in the reservoirs.

9.6.2.2 Policies Relating to Source and Technology Selection

The policy regarding the selection of technology for water supply should be driven by resource sustainability and source sustainability considerations rather than the consideration of decentralized management. Ideally, only those sources wherein the physical allocation of water for domestic water supply is technically feasible and which is not likely to face competition from other sectors of water use need to be chosen. The schemes that tap water from flowing streams and rivers should be resorted to only when such sources are highly dependable in terms of quantity and quality across seasons. Further, groundwater-based schemes can be resorted to in alluvial areas where the chances of seasonal depletion of the aquifers do not exist and the resource is free from contaminants.

9.6.2.3 Policies Relating to Institutional Models

Since decentralization is given thrust in rural water supply, the current tendency of state water supply agencies is to go for schemes that the Panchayat water supply and sanitation committees can manage. However, village, community-based organizations are created today using a stereotypical approach focusing on representation from community members, but with no major support for technical issues, and monitoring and finances (Kumar et al. 2021). Recent research shows that for community management models to be successful and sustainable, a number of internal and external factors have to act in tandem such as extensive and long-term support for local community institutions; and broader socio-economic development of the community. Instead, the nature of technology chosen for water supply based on the consideration of the sustainability of the resource and source should guide the institutional choice (Hutchings et al. 2015).

Water supply schemes based on large reservoirs can offer economies of scale, only when a large number of villages are covered under a single scheme. This would obviously offer a lot of challenges for design and operation from the point of view of hydraulic engineering. This is because such systems would require sophisticated high-capacity pumping machinery, long-distance water transport, piped water distribution, flow regulators, pressure control devices, and online real-time monitoring using systems such as SCADA. Professional centralized institutions with required technical know-how and expertise and financial resources are required to design, build and run such technically sophisticated schemes. More importantly, the institution, which designs the scheme, should ideally run it for efficient management. The Gujarat Water Supply and Sewerage Board (GWSSB), which runs a large regional water supply scheme based on bulk water supply from Narmada canal, involving several thousand kilometre of pipe networks and covering thousands of villages and towns, is a good example (Biswas-Tortajada 2014).

9.6.2.4 Policies and Norms on Water Supply Levels

There are no clear-cut policies and norms relating to the level of water supply to be maintained in rural water supplies in the country. The water supply for the state-funded schemes is generally decided on the basis of a standard norm of 40 lpcd, whereas in some of the externally funded schemes, the supply norm ranges from 55 to 70 lpcd. These norms are not arrived at based on any scientific assessment of the actual water requirement of the rural households for meeting the domestic water needs for maintaining good health and hygiene and environmental sanitation. If improved latrines are to be adopted by rural families, then the per capita domestic water requirements would increase significantly. Additional water would be required to take care of the productive water needs such as livestock rearing, backyard cultivation, and kitchen gardens (Kumar et al. 2021). As noted by Bassi et al. (2021), the policy on water supply levels should take into account socio-economic and climatic factors while working out the domestic water requirement for different regions. At the same time, it should be an ‘enabling’ framework that allows the line agencies to work out the per capita supply based on the criteria set by the local self-governing institutions.

9.6.2.5 Policies on Pricing and Subsidies for Water Supply

Ideally, for ensuring the financial viability of the schemes, the pricing of water for domestic water supply should be driven by marginal cost considerations. Here, the marginal cost would include the following: (i) the long-term marginal cost of infrastructure for production and supply; (ii) the environmental cost of degradation of water resources due to diversion from the natural system and pollution of water bodies due to wastewater return flows; and, (iii) the opportunity cost of depletion of the water. If we go by this norm, the cost of water supply would keep varying widely between regions, with very high costs in naturally water-scarce regions and comparatively low costs in water abundant and water surplus regions (Kumar 2014). Secondly, for many water supply systems, particularly those in water-scarce regions, the long-term marginal cost of water production supply would work out to be so enormous that most people in rural areas would find it un-affordable. Since what is being achieved through the provision of water supply for basic needs is a ‘social good’, with positive welfare effects of individual household water security (UNICEF/IRAP 2013), it is desirable to introduce subsidies in order to ensure that every household is able to access water to meet the basic requirements.

Hence, a uniform price for water across regions, without taking into cognizance the socio-economic realities and the cost of obtaining water, would be irrational. The policy for water pricing should be such that the actual prices are decided by considerations of affordability as well as the ‘long-term marginal cost’. Adoption of such a policy would allow flexibility in fixing water charges. It means that in some regions that are water-scarce, the extent of subsidy required would be very high, whereas it would be quite low in regions that are water abundant and where the communities are rich (Kumar et al. 2021).

9.6.2.6 Investment Policies

There is a need for life cycle cost analysis (LCCA) in choosing the best alternatives in water supply systems. The investment policies relating to water supplies need to take into account the concerns of source adequacy, source dependability, and long-term sustainability of the sources and cannot be merely driven by size of the one-time capital investment required per capita, or the O&M requirements per capita (Reddy et al. 2012). Taking into account considerations such as opportunity cost of inadequate supplies, poor dependability of the source or the permanent failure of the source before the design life in costing of various technical models of water supply would encourage investments into expensive (capital intensive) options, which are cost-effective in the long run. The analysis presented in Chap. 4 has shown that the cost of supplying water to the villages through tankers during the summer months, when the bore wells run dry, can be considerably high.

9.7 Conclusions

The groundwater-based schemes that dominate India's rural water are highly susceptible to risk induced by extreme climatic conditions such as droughts and floods, and resource over-exploitation. In some regions having very high to excessively high rainfall and sub-humid to humid climatic conditions and where groundwater is plenty, the resource is susceptible to microbial and nitrate pollution. In the past, the design of such schemes has not taken into consideration how various natural and socio-economic factors cause disruptions in water supply, primarily the droughts that reduce recharge and increase the demand for irrigation water, and the uncontrolled withdrawal of underground water for irrigation that depletes the aquifers. This has led to problems of 'slip-back', with the drying up of village drinking water sources. Lack of proper surveillance of drinking water sources for water quality management is also an important concern.

In this chapter, we have discussed the strategies for improving the sustainability of rural water supply systems. While there are several interventions that can reduce the risks posed to drinking water schemes, supplying treated water from large reservoirs that tap flows from rich catchments is a critical intervention. Hence, the investigations in search of the strategies have considered the surface water and groundwater availability across space (in different river basins), its inter-annual and inter-seasonal variability, the current pressure on the resources from other sectors, the chemical quality of groundwater and topography.

Sourcing water from large reservoirs fed by rich catchments enables serving several villages from a single source and therefore brings in economies of scale. Hence, it enables the installation of capital-intensive water treatment plants which will not be feasible for individual village schemes. The uniqueness of this strategy is that once the households are assured of treated water throughout the year, they will go for individual household tap connections, further reducing the risks. This will be achieved through the following: (i) local water resources development based on small dams (in the hilly regions of the north-east, and in the Western Ghats region) and lifting the water to higher elevations and by gravity pipes to lower elevations; (ii) building large reservoirs in river basins that still have un-utilized water resources and building distribution infrastructure; (iii) reallocating some water from large reservoirs in the water-scarce regions, and building distribution infrastructure; (iv) inter-basin water transfer projects that benefit many regions with reservoirs in the water-rich basins and transporting to far-off areas through pipeline networks and canals.

Several inter-basin water transfer projects are either under implementation or under consideration in India (Kumar 2018). As experience shows, annual import of large volume of surface water to the naturally water-scarce regions and expansion of gravity irrigation through inter-basin water transfer and large, conventional water development projects not only reduce the pressure on the limited groundwater resources for irrigation, they will also induce recharge to the shallow aquifers. This in turn can have positive effects on drinking water supply systems that tap groundwater, in terms of reducing their exposure to the effects of droughts on natural replenishment (Jagadeesan and Kumar 2015). However, for sustainable investments in large water infrastructure projects for drinking water security, institutional and policy reforms are needed.

References

Bassi N, Kabir Y, Ghodke A (2021) Planning of rural water supply systems: role of climatic factors and other considerations. In: Kumar MD, Kabir Y, Hemani R, Bassi N (eds) Management of irrigation and water supply under climatic extremes: empirical analysis and policy lessons from India. Springer Nature, Switzerland, pp 161–177

Biswas-Tortajada A (2014) Gujarat state-wide water supply grid: a step towards water security. Int J Water Resour Dev 30(1):78–90

Central Ground Water Board (2020) Ground water year book 2019–20. Central Ground Water Board, Ministry of Jal Shakti, Faridabad

Central Water Commission (2017) Reassessment of water availability in India using space inputs. Basin Planning and Management Organization, Central Water Commission, New Delhi

Government of India (1999) Integrated water resource development a plan for action, report of the National Commission on Integrated Water Resources Development, Volume I. Ministry of Water Resources., Govt. of India, New Delhi

Guhathakurta P, Sreejith OP, Menon PA (2011) Impact of climate change on extreme rainfall events and flood risk in India. J Earth Syst Sci 120(3):359–373

Hemani R, Bassi N, Kumar MD, Chandra U (2021) Mapping climate-induced risk for water supply, sanitation and hygiene in Rajasthan. In: Kumar MD, Kabir Y, Hemani R, Bassi N (eds) Management of irrigation and water supply under climatic extremes: empirical analysis and policy lessons from India. Springer Nature, Switzerland, pp 241–285

Hutchings P, Chan MY, Cuadrado L, Ezbakhe F, Mesa B, Tamekawa C, Franceys R (2015) A systematic review of success factors in the community management of rural water supplies over the past 30 years. Water Policy 17(5):963–983

IRAP, GSDA, UNICEF (2013) Multiple-use water services to reduce poverty and vulnerability to climate variability and change: a collaborative action research project in Maharashtra, India. Institute for Resource Analysis and Policy, Hyderabad

Jagadeesan S, Kumar MD (2015) The sardar sarovar project: assessing economic and social impacts. Sage Publications, New Delhi

Jha BM, Sinha SK (2009) Towards better management of ground water resources in India. Bhu-Jal News 24(4):1–20

Kabir Y, Niranjan V, Bassi N, Kumar MD (2016) Multiple water needs of rural households: studies from three agro-ecologies in Maharashtra. In: Kumar MD, Kabir Y, James AJ (eds) Rural water systems for multiple uses and livelihood security. Elsevier, Netherlands, pp 49–68

Kraft GJ, Stites W (2003) Nitrate impacts on groundwater from irrigated-vegetable systems in a humid north-central US sand plain. Agr Ecosyst Environ 100(1):63–74

Krishnamurthy V, Shukla J (2000) Intraseasonal and interannual variability of rainfall over India. J Clim 13(24):4366–4377

Kumar MD (2014) Thirsty cities: how Indian cities can meet their water needs. Oxford University Press, New Delhi

Kumar MD (2018) Future water management: myths in Indian agriculture. In: Biswas AK, Tortajada C, Rohner P (eds) Assessing global water megatrends. Springer, Singapore, pp 187–209
Kumar P, Shukla BP, Sharma S, Kishtawal CM, Pal PK (2016) A high-resolution simulation of catastrophic rainfall over Uttarakhand, India. Nat Hazards 80(2):1119–1134
Kumar MD, Bassi N, Hemani R, Kabir Y (2021) Managing climate-induces water stress across the agro-ecological regions of India: options and strategies. In: Kumar MD, Kabir Y, Hemani R, Bassi N (eds) Management of irrigation and water supply under climatic extremes: empirical analysis and policy lessons from India. Springer Nature, Switzerland, pp 313–354
Kumar MD, Sivamohan MVK, Narayanamoorthy A (2012) The food security challenge of the food-landwater nexus in India. Food Security 4(4):539–556
Narmada Water Disputes Tribunal (NWDT) (1979) The report of the Narmada Water Disputes Tribunal Vol I, Narmada Water Disputes Tribunal, Government of India, New Delhi
National Institute of Hydrology (1999) Rainfall-runoff modelling of Western Ghat region of Karnataka. National Institute of Hydrology, Roorkee, India.
Pal I, Al-Tabbaa A (2011) Regional changes of the severities of meteorological droughts and floods in India. J Geogr Sci 21(2):195–206
Pisharoty PR (1990) Characteristics of Indian rainfall (Monograph). Physical Research Laboratories, Ahmedabad
Prokop P, Walanus A (2017) Impact of the Darjeeling-Bhutan Himalayan front on rainfall hazard pattern. Nat Hazards 89(1):387–404
Rajeevan M, Bhate J, Jaswal AK (2008) Analysis of variability and trends of extreme rainfall events over India using 104 years of gridded daily rainfall data. Geophys Res Lett 35(18)
Rao BB, Sandeep VM, Rao VUM, Venkateswarlu B (2012) Potential evapotranspiration estimation for Indian conditions: improving accuracy through calibration coefficients. Technical Bulletin No 1/2012, All India Co-ordinated Research Project on Agrometeorology, Central Research Institute for Dryland Agriculture, Hyderabad
Rawat PK, Tiwari PC, Pant CC, Sharma AK, Pant PD (2011) Climate change and its geo-hydrological impacts on mountainous terrain: a case study through remote sensing and GIS modeling. E-Int Sci Res J 3(1):51–69
Reddy VR, Snehalatha M, Venkataswamy M (2012) Costs and service per technology in rural water supply: how efficient are multi village schemes? WASHCost (India), Working Paper 18, Centre for Economic and Social Studies, Hyderabad
UNICEF, IRAP (2017) Capacity building for planning of climate-resilient WASH services in rural Maharashtra. UNICEF, Mumbai
Winter TC, Harvey JW, Franke OL, Alley WM (1998) Ground water and surface water a single resource. US Geological Survey Circular 1139, Denver, Colorado

Chapter 10
Rural Drinking Water Security in India: The Challenge of Piping Water to Every Household by 2024

10.1 Introduction

On 15 June 2019, the Prime Minister of India, Shri Narendra Modi, announced that piped water would reach all homes in India's countryside by 2024. This in no way meant that the rural people of India are not able to access clean water now. What the Prime Minister wanted to see is that every household can access clean water in the dwelling premise or the 'Ease of living' in villages. Had it been the case that most people are not able to access clean water in Indian villages, the incidence of waterborne diseases in many regions of the country would have been many times greater. Instead, what was meant is that the physical distance to the water source has to be reduced. Often women, children, and sometimes even men walk long distances to fetch clean water from public sources. There is a huge socio-economic benefit that can be derived from providing clean water within the dwelling premise, which is hardly ever evaluated. It is not merely the public health benefit of drinking clean water alone. This is where the new announcement from the Hon. PM assumes enormous significance. Let us see what these benefits are, how they are going to be accrued, and what it takes to provide them.

In fact, over the past few decades, we had made huge progress in covering rural areas of the country with public water supply (PWS) systems, most of it supplying water through pipes. In most states, it is as high as 85–90%. In a small percentage of the villages and hamlets, the communities depend either on 'other improved sources of water' supply (handpumps, bore wells, etc.) or water from open wells, tanks, and ponds that can be unsafe for human health. Overall, a very large percentage of the households are able to access a supply of clean water through pipes—which can be near to the dwelling or away from the dwelling. A small percentage of the households accesses 'tap water' in their dwelling (about 18% as per the latest estimates), which by definition is treated water supplied through pipes in the dwelling premise.

Targeting for tap water supply for all households in rural areas is indeed very ambitious. First of all, the water has to be treated. Second: it has to be piped to the dwelling. Now in India, it is well-known fact that most of the rural water supply

M. Dinesh Kumar et al., *Drinking Water Security in Rural India*, Water Resources Development and Management, https://doi.org/10.1007/978-981-16-9198-0_10

schemes are based on underground sources such as tube wells, open wells, and handpumps.

As the official agencies do not take cognizance of the life cycle cost of the scheme (Reddy et al. 2012), the obvious preference is for groundwater-based schemes over surface water schemes because of the lower capital investment for the former. But as our analysis in previous chapters has shown, the underground water in vast regions of India suffers from one contamination problem or the other—either an excessively high level of fluoride or salinity or nitrate or chloride or iron or arsenic. Siting of wells for drinking water supply is done in such a way that the concentration of the above parameters in the raw water is within permissible limits in most situations. The source can therefore normally provide safe water, but it is 'untreated'. When a piped water supply scheme is developed around underground sources in a village, the only treatment that this water is subject to is disinfection (to protect it from contamination during conveyance), as the raw water is declared free of chemical contaminants. But at any point in time, the water can turn unsafe, depending on the geogenic processes that occur underground.

10.2 The Benefits and Impacts of 'Tap Water'

There are instances when people in villages invest money for household water connections and enjoy 'tap water'. The proportion is around 18% for villages. But in most such cases, the source water is surface water from large reservoirs and not groundwater from tube wells, bore wells, and handpumps. The reason is that in the case of surface water, the people are almost sure that the scheme will be able to supply water free of chemical contaminants throughout its life, as the source can be protected. The second reason is that people are sure of the reliability of the source that it can supply water during the summer months as well. Hence, these are the following three benefits of accessing tap water generally, though the last two are hidden:

1. Substantial amount of time is saved in fetching water when water is available within the dwelling;
2. The communities will have much greater assurance that the water is safe and do not have to purchase costly RO water supplied by private vendors due to the fear of getting exposed to chemically and microbially contaminated water; and
3. The source is highly dependable, unlike in the case of groundwater where the sources (wells and handpumps) go dry during summer months owing to pressure from the agriculture sector, and the village communities have to depend on government tanker water supply. Thus, the socio-economic benefits of 'tap water' are large.

Another indirect benefit is that with access to piped water within the dwelling rural households are likely to go for improved toilets and use them properly. Though many millions have constructed toilets in their dwelling premises during the past

7 years under the Swachh Bharat Mission (SBM) (107.1 million toilets in rural areas and 6.6 million toilets in urban areas built since October 2014) (SBM-G; SBM-U), most of them are unlikely to use them due to inadequate access to running water. These toilets, which were built with a lot of public money spent on capital subsidy, would soon go into disuse. This problem can be solved through the provision of 'tap water', which is defined as the supply of treated water to individual households through tap connections within the dwelling.

Though in many rural areas of India (Gujarat, Rajasthan, Telangana, Karnataka, and AP), groundwater is being treated for removal of chemical contaminants such as fluorides and salts (when its concentration exceeds permissible levels and the surface water sources are not available in the vicinity) using defluorination plants and reverse osmosis (RO) plants, respectively, under various institutional arrangements (source: based on Kumar 2017; Reddy 2018), the treated water is available only for drinking and cooking purposes. Plants for large-scale treatment of chemically contaminated water would lead to large volumes of reject water whose disposal would pose huge environmental problems while raising the treatment cost to prohibitively high levels (Rs. 30–40/k L of water). So, in such situations, even if piped water supply is available to every household for other domestic uses, the communities will still need to come to the plants for procuring water for drinking and cooking purposes.

The most sustainable option, therefore, is to go for reservoir-based schemes that are large enough to supply good quality water to hundreds (or sometimes thousands) of villages to meet all domestic needs (including washing, bathing, toilet use apart from drinking, and cooking). In this case, it is possible to treat the raw water at low costs (as the sediments, biological matter, and microbial contaminants that are common can be removed by simple filtration) and economies of scale can be achieved. In Gujarat, one gigantic water supply scheme based on the Narmada canal supplies water to several thousands of villages and hundreds of towns in the parched areas of Saurashtra and Kachchh (Jagadeesan and Kumar 2015). Up till March 2019, 8,911 villages and 165 towns have been benefitted by drinking/domestic water supplies from Narmada through pipelines, and the Narmada Master Plan is aimed at covering 9,490 villages and 173 towns (SSNNL) (source: https://sardarsarovardam.org/water-supply-policy.aspx).

Unfortunately, one of the biggest mistakes our policymakers in the drinking water sector had made during the past few decades is the overemphasis they had given on groundwater-based, mini schemes in the pre-text of being able to develop the scheme quickly hand over their management to the village Panchayats. This shift had happened because of the willingness communities show for taking over the operation and maintenance of such simple schemes. In the case of large regional water supply schemes, this willingness is absent due to the complex nature of the technical system which has to be run and maintained. The result is that schemes have a very short life and 'slippage' is very high (Reddy et al. 2010). Villages having 5–6 such schemes, yet having serious water shortages during summer months are common.

10.3 Link Between Quality of Water Supply and Household Tap Connection

If rural water security is to be achieved and if the full benefits (including many socio-economic benefits) of having clean water are to be realized, it is important that we invest in large reservoir-based schemes. Only they can provide clean, chemical-free water at low costs that too with high dependability. Only then the communities would show the willingness to invest in household water connections. This was illustrated by a study done in Maharashtra in 2012. The study used block-level data on the number of households, from each of the three distinct categories of access (viz. water within the premise, water source near the premise, or water source away from the premise), using treated piped water supply to see whether there was any trend in preference of the rural households via-a-vis type of domestic water supply they would like to enjoy. Treated piped water supply was chosen because it is the most ideal source of water people like to have in villages in terms of quality and ease. Since people can also change the degree of access on the basis of their payment power, the type of domestic water supply would also indicate the willingness to pay (UNICEF/IRAP 2013).

Analysis involved comparing the total number of HHs depending on treated tap water from each block in each category of 'water access', as a percentage of the total HHs in the block. The three categories of access were 'within the dwelling premise', 'near the dwelling', and 'away from the dwelling premise'. The results of the analysis showed that a much larger percentage of HHs obtain treated tap water in their dwelling (64%) than those who get it near the dwelling (29%). The percentage difference between category 1 and category 3 was even larger (57%) as compared to the percentage difference between category 1 and category 2 (35%). The percentage HHs accessing treated tap water in their dwelling was consistently higher than counterparts with distant sources across the blocks (Fig. 10.1). Getting

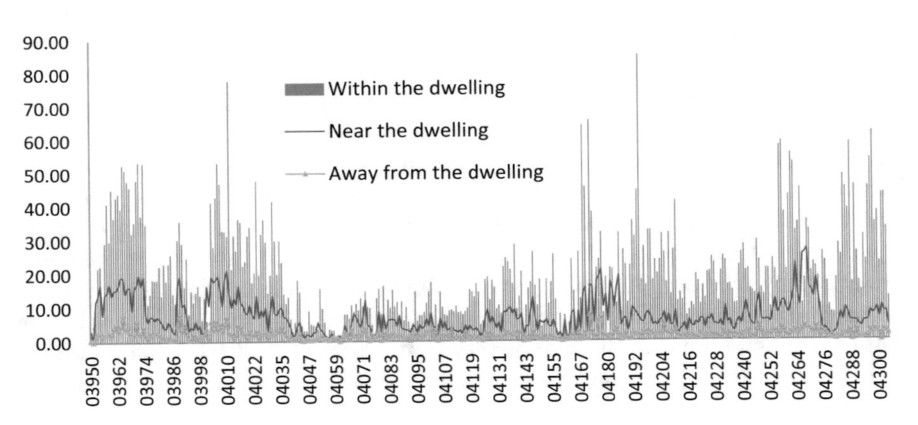

Fig. 10.1 Differential access to water with changing quality of service: treated tap water. *Source* UNICEF/IRAP (2013)

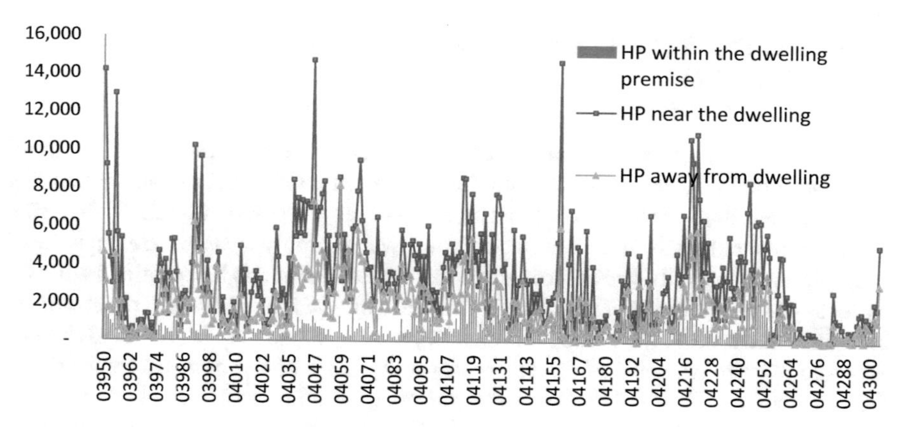

Fig. 10.2 Proportion of population depending on HPs, with varying degree of access. *Source* UNICEF/IRAP (2013)

connection for treated tap water within the dwelling premise is much more expensive than getting it near the dwelling premise and the cheapest source of 'tap water' is that which is obtained from distant common sources such as the common stand posts. This trend shows that when water is of good quality and there is ease in collecting it, HHs show greater willingness to pay for water supply services.

A similar analysis was carried out for handpumps in each of these blocks. In the case of handpumps, nearly 15.7% of the households depend on this improved source of water supply in the entire state. The total number of households that depend on handpumps was 20.43 lac. But, only 13.8% of them had HPs in their dwelling premises. Since these handpumps are located within the dwelling premise, one can assume that these are privately owned. More than half of them (56.2%) depend on handpumps near the dwelling premise. In all, around 30% of the households depend on HPs located away from the dwelling premise. Therefore, in total, in around 86 per cent of the cases, the HPs must be publicly owned as they are either located near the dwelling premise or away from the dwelling premise. Figure 10.2 shows the total number of households that have access to handpumps across different Tehsils of Maharashtra, within the dwelling, near the dwelling, and away from the dwelling, respectively. It can be seen from the figure that in a majority of the Tehsils, more households depend on distant HPs as compared to those who depend on HPs located within the dwelling or near the dwelling. What is clear from the comparative analysis of data on the type of water access for 'treated tap water' and 'handpump water' is that when water is untreated, and not supplied through taps, rural people do not like to invest in going for domestic connection. Conversely, when the water is of good quality, most people like to invest money for accessing it within their dwelling (UNICEF/IRAP 2013).

10.4 Realizing the Vision of 'Tap Water at Every Household'

If the vision of piping water to every Indian household is to become a reality, India has to start building new infrastructure in the form of large reservoirs in some cases and water distribution systems in most other cases (as discussed in Chap. 9). In terms of geographical extent, the areas where rural water supply systems are capable of supplying good quality water throughout the year and that can be sustained entirely by local groundwater is not very large. The criticality of having surface reservoirs as the source of drinking water supply is that they can be planned vis-a-vis the catchment area and storage capacity of the reservoir on the basis of the demand for water in the 'service area' (the villages) and the run-off generation possible (yield) per unit area of the catchment with a high degree of dependability and physical allocation of water from the reservoir in volumetric terms for the earmarked uses is possible unlike in the case of groundwater reserves.

In the rest of the areas, investing more money on groundwater-based schemes will only result in increased vulnerability of the communities to problems associated with lack of water of sufficient quality and quantity, as the government is not able to enforce any control on groundwater abstraction by farmers. If we really want to see the impact of providing clean water to the rural communities, the best illustrative example are Ganganagar and Jaisalmer districts of western Rajasthan—one of the most arid and water-scarce regions in the world—whose scores in human development (0.76 and 0.61, respectively) are two of the highest in the state (Hemani et al. 2021). The face of the two districts had changed because of the water imported from the Sutlej River through the Indira Gandhi Nahar Project (IGNP) canal, with every rural household getting plenty of clean water in their dwelling for all household uses, along with water for food production.

But achieving this target is not going to be easy. This ambitious scheme will consume large amounts of financial resources from the Nation's exchequer. New schemes will have to be built replacing the old one, since simply extending the pipelines of the existing PWS schemes that tap unreliable sources (groundwater, tanks, etc.) will not make any sense to the households. New sources will have to be developed, along with creating distribution and delivery infrastructure for the water stored in the reservoirs.

If we assume a per capita investment of Rs. 10,000 (US$150 per capita), this would roughly cost the government a whopping 90 billion dollars in the five years from 2020 to 2024 (for a population of 645 million people in the rural areas who do not get treated water in their dwelling). Such an infrastructural investment will have a life of 50 years or so. An alternative approach will be to look at the annualized cost per cubic metre of the water to be supplied through infrastructure such as reservoirs and water distribution pipelines. If we consider an average annualized unit cost of Rs. 7.4/m^3 for such an infrastructure (source: based on calculations presented in Chap. 6) and an average per capita average water supply of 100 L per day, the investment required for a population of 645 million would be US$2.32 billion on an annual basis. But this will dramatically change life in rural areas: women will have more leisure; there would be better adoption and use of improved toilets and therefore less occurrence of diseases; and there would be fewer cases of child malnutrition. The highly exploitative drinking water markets, which have come up in many areas due to state failure in provisioning adequate quantities of good quality water for drinking throughout the year, will be eliminated. Probably, there would be fewer cases of school dropouts too.

In terms of economic benefits, this permanent investment would help save the money spent by state governments on supplying water to villages through tankers, done especially during the peak summer months. If we assume that a mere 20% of the 645 million people in the rural areas (who are supplied drinking water from groundwater-based schemes) are to be served through tanker water supply during summer (3 months) and these people manage their water needs with just 40 lpcd (supplied through tankers), and the average cost of tanker water supply (to the government) is to be Rs. 250/m^3, the total cost of supplying water to meet the summer deficit comes to Rs. 11,600 crore per annum or US$1.56 billion. This can be considered as one direct benefit. In addition, there would be benefits of time and labour saving for the people, especially women who have to stand in queues to fetch water from the tankers on a daily basis. If we consider the economic value of this gain to be Rs. 40 (US$0.60) for a daily time saving of one hour, the money equivalent of this comes to Rs. 9,288 crore (US$1.29 billion) per annum (for 25.8 million families, for 90 days a year). The total annual economic benefit works out to be US$2.85 billion, against an annual cost of US$2.2 billion. The prevention of welfare loss due to adverse health outcomes caused by people not being able to meet all their household water needs, resulting from a reduced supply of good quality water, whose quantification is attempted here, is an additional indirect benefit. In sum, the financial investments for improving rural drinking water security in India will be large, but the positive social and economic outcomes will be much larger.

References

Hemani R, Bassi N, Kumar MD, Chandra U (2021) Mapping climate-induced risk for water supply, sanitation and hygiene in Rajasthan. In: Kumar MD, Kabir Y, Hemani R, Bassi N (eds) Management of irrigation and drinking water supply under climate extremes: empirical analysis and policy lessons from India. Springer, Singapore, pp 241–286

Jagadeesan S, Kumar MD (2015) The sardar sarovar project: assessing economic and social impacts. Sage Publications, New Delhi

Kumar MD (2017) Market analysis: desalinated water for irrigation and domestic use in India, prepared for securing water for food: A grand challenge for development in the Center for Development Innovation, US Global Development Lab. DAI Professional Management Services, US

Reddy VR (2018) Techno-institutional models for managing water quality in rural areas: case studies from Andhra Pradesh, India. Int J Water Resour Dev 34(1):97–115

Reddy VR, Rammohan Rao MS, Venkataswamy M (2010) 'Slippage': the bane of drinking water and sanitation sector (a study of extent and causes in rural Andhra Pradesh). WASHCost India-CESS Working Paper, Hyderabad, India

Reddy VR, Jayakumar V, Venkataswamy M, Snehalatha M, Batchelor C (2012) Life-cycle costs approach (LCCA) for sustainable water service delivery: a study in rural Andhra Pradesh, India. J Water Supply Sanit Hygiene Dev 2(4):279–290

Sardar Sarovar Narmada Nigam Limited. https://sardarsarovardam.org/water-supply-policy.aspx. Accessed 21 Sept 2021

Swachh Bharat Mission-Grameen. https://sbm.gov.in/sbmReport/home.aspx. Accessed 21 Sept 2021

Swachh Bharat Mission-Urban. http://swachhbharaturban.gov.in/. Accessed 21 Sept 2021

UNICEF, Institute for Resource Analysis and Policy (2013) Promoting sustainable water supply and sanitation in rural Maharashtra: Institutional and policy regimes, report of the study done in association with the Water Supply and Sanitation Department, Govt. of Maharashtra, Mumbai. Institute for Resource Analysis and Policy, Hyderabad

Index

M. Dinesh Kumar et al., *Drinking Water Security in Rural India*, Water Resources Development and Management, https://doi.org/10.1007/978-981-16-9198-0

Y

Printed in the United States
by Baker & Taylor Publisher Services